Left	Right
Louise	Ri
Positive	Neg
bLack	W
bLue	Red
Vertical	Horizontal
Female	Male
Green	
Gray	

Symbol	Description	Page
$G = \left\{ \mathcal{G}^L \mid \mathcal{G}^R \right\}$	Definition of a game	37, 76
G^L and G^R	Typical left option and right option	37
$\{A \parallel B \mid C, D\}$	Notation for a game	37
$G + H$	$\left\{ \mathcal{G}^L + H,\ G + \mathcal{H}^L \mid \mathcal{G}^R + H,\ G + \mathcal{H}^R \right\}$	78
$-G$	$\left\{ -\mathcal{G}^R \mid -\mathcal{G}^L \right\}$ (negative)	79
$G : H$	Ordinal (or sequential) sum	239
$G \cong H$	identical game trees (congruence)	76
$G = H$	game equivalence	80
$\geq, \leq, >, <$	comparing games	83
$G \parallel H$	$G \ngeq H$ and $H \ngeq G$ (incomparable, confused with)	84
$G \rhd H$	$G \nleq H$ (greater than or incomparable)	84
$G \lhd H$	$G \ngeq H$ (less than or incomparable)	84
$G \sim_{\star} H$	far star equivalence	216
$G \sim_{\epsilon} H$	infinitesimally shifted	234
n	integers	106
$m/2^j$	numbers or dyadic rationals	108
$\uparrow$ and $\downarrow$	$\{0 \mid *\}$ "up" and $\{* \mid 0\}$ "down"	119
$\Uparrow$ and $\Downarrow$	$\uparrow + \uparrow$ "double-up" and $\downarrow + \downarrow$ "double-down"	119
$*$	$\{0 \mid 0\}$ "star"	??
$*n$	$\{0, *, *2, \ldots \mid 0, *, *2, \ldots\}$ "star-n" (nimbers)	156
$\mathbf{+}_G;\ \mathbf{-}_G$	$0 \parallel 0 \mid -G$ "tiny-G" and its negative "miny-G"	127
$\mathfrak{D}$	loony	22
$\star$	far star	217
$\mathbf{LS}(G),\ \mathbf{RS}(G)$	(adorned) left and right stops	(183) 117
$\mathcal{G}_n$ and g_n	games born by day n and their number	139
$\vee$ and $\wedge$	join and meet	148
$n \cdot G$	$\overbrace{G + G + G + \cdots + G}^{n}$	122
$G \cdot U$	Norton product	197
G^n	"G-nth"	210
$.213$	uptimal notation	210
$G^{\rightarrow n}$	$G^1 + G^2 + \cdots + G^n$	211
$\mathrm{AW}(G)$	atomic weight	219

Lessons in Play

Second Edition

Lessons in Play

An Introduction to
Combinatorial Game Theory

Second Edition

Michael H. Albert
Richard J. Nowakowski
David Wolfe

CRC Press
Taylor & Francis Group
Boca Raton London New York

CRC Press is an imprint of the
Taylor & Francis Group, an **informa** business

AN A K PETERS BOOK

CRC Press
Taylor & Francis Group
6000 Broken Sound Parkway NW, Suite 300
Boca Raton, FL 33487-2742

First issued in paperback 2022

ISBN 13: 978-1-03-247566-0 (pbk)
ISBN 13: 978-1-4822-4303-1 (hbk)

Library of Congress Cataloging-in-Publication Data

Names: Albert, Michael H., author. | Nowakowski, Richard J., author. | Wolfe, David,
1964- author.
Title: Lessons in play : an introduction to combinatorial game theory / Michael Albert,
Richard Nowakowski, David Wolfe.
Description: Second edition. | Boca Raton, Florida : CRC Press, [2019] |
Includes bibliographical references and index.
Identifiers: LCCN 2018060986| ISBN 9781482243031 (hardback : alk. paper) |
ISBN 9781482243048 (ebook : alk. paper)
Subjects: LCSH: Game theory--Textbooks. | Combinatorial analysis--Textbooks.
Classification: LCC QA269 .A425 2019 | DDC 519.3--dc23
LC record available at https://lccn.loc.gov/2018060986

Visit the Taylor & Francis Web site at
http://www.taylorandfrancis.com

and the CRC Press Web site at
http://www.crcpress.com

To Richard K. Guy, a gentleman and a mathematician

Contents

Contents   ix

Preface

It should be noted that children's games are
not merely games. One should regard them
as their most serious activities.

Michel Eyquem de Montaigne

Herein we study games of pure strategy, in which there are only two players[1] who alternate moves, without using dice, cards, or other random devices, and where the players have perfect information about the current state of the game. Familiar games of this type include TIC TAC TOE, DOTS & BOXES, CHECKERS, and CHESS. Obviously, card games such as GIN RUMMY and dice games such as BACKGAMMON are not of this type. The game of BATTLESHIP has alternate play and no chance elements, but fails to include perfect information — in fact, that's rather the point of BATTLESHIP. The games we study have been dubbed *combinatorial games* to distinguish them from the games usually found under the heading of *game theory*, which are games that arise in economics and biology.

For most of history, the mathematical study of these games consisted largely of separate analyses of extremely simple games. This was true up until the 1930s when the Sprague-Grundy theory provided the beginnings of a mathematical foundation for a more general study of games. In the 1970s, the twin tomes *On Numbers and Games* by Conway and *Winning Ways* by Berlekamp, Conway, and Guy established and publicized a complete and deep theory, which can be deployed to analyze countless games. One cornerstone of the theory is the notion of a disjunctive sum of games, introduced by John Conway for normal-play games. This scheme is particularly useful for games that split naturally into components. *On Numbers and Games* describes these mathematical ideas at a sophisticated level. *Winning Ways* develops these

---

[1]In 1972, Conway's first words to one of the authors, who was an undergraduate at the time, was "What's $1 + 1 + 1$?" alluding to three-player games. This question has still not been satisfactorily answered.

ideas, and many more, through playing games with the aid of many a pun and witticism. Both books have a tremendous number of ideas, and we acknowledge our debt to the books and to the authors for their kind words and teachings throughout our careers.

The goal of our book is less grand in scale than either of the two tomes. We aim to provide a guide to the evaluation scheme for normal-play, two-player, finite games. The guide has two threads, the theory and the applications.

The theory is accessible to any student who has a smattering of general algebra and discrete mathematics. Generally, this means a third-year college student, but any good high school student should be able to follow the development with a little help. We have attempted to be as complete as possible, though some proofs in the latter chapters have been omitted, because the theory is more complex or is still in the process of being developed. Indeed, in the last few months of writing the first edition, Conway prevailed on us to change some notation for a class of all-small games. This *uptimal* notation turned out to be very useful, and it makes its debut in this book.

We have liberally laced the theory with examples of actual games, exercises and problems. One way to understand a game is to have someone explain it to you; a better way is to think about it while pushing some pieces around; and the best way is to play it against an opponent. Completely solving a game is generally hard, so we often present solutions to only some of the positions that occur within a game. The authors invented more games than they solved during the writing of this book. While many found their way into the book, most of these games never made it to the rulesets found at the end. A challenge for you, the reader of our missive, and as a test of your understanding, is to create and solve your own games as you progress through the chapters.

Since the first appearance of *On Numbers and Games* and *Winning Ways,* there have been several conferences specifically on combinatorial games. The subject has moved forward and we present some of these developments. However, the interested reader will need to read further afield to find the theories of loopy games, misère-play games, other (non-disjunctive) sums of games, and the computer science approach to games. The proceedings of these conferences [Guy91, Now96, Now02, AN07, Now15, FN04] would be good places to start.

Organization of the Book

The main idea of the part of the theory of combinatorial games covered in this book is that it is possible to assign values to games. These values, which are not simply numbers, can be used to replace the actual games when deciding who wins and what the winning strategies might be.

Each chapter has a prelude that includes problems for the student to use as a warm-up for the mathematics to be found in the following chapter. The prelude also contains guidance to the instructor for how one can wisely deviate from the material covered in the chapter.

Exercises are sprinkled throughout each chapter. These are intended to reinforce, and check the understanding of, the preceding material. Ideally then, a student should try every exercise as it is encountered. However, there should be no shame associated with consulting the solutions to the exercises found at the back of the book if one or more of them should prove to be intractable. If that still fails to clear matters up satisfactorily, then it may be time to consult a *games guru*.

Chapter 0 introduces basic definitions and loosely defines that portion of game theory which we address in the book. Chapter 1 covers some general strategies for playing or analyzing games and is recommended for those who have not played many games. Others can safely skim the chapter and review sections on an as-needed basis while reading the body of the work. Chapters 2, 4, and 5 contain the core of the general mathematical theory. Chapter 2 introduces the first main goal of the theory, that being to determine a game's *outcome class* or who should win from any position. Curiously, a great deal of the structure of some games can be understood solely by looking at outcome classes. Chapter 3 motivates the direction that the theory takes next. Chapters 4, 5, and 6 then develop this theory (i.e., assigning values and the consequences of these values).

Chapters 7, 8, and 9 look at specific parts of the universe of combinatorial games, and as a result, these are a little more challenging but also more concrete since they are tied more closely to actual games. Chapter 7 takes an in-depth look at *impartial* games. The study of these games pre-dates the full theory. We place them in the new context and show some of the new classes of games under present study.

Chapters 8 through 10 provide techniques for identifying and exploiting the most significant information about a game when a complete analysis might be complex and therefore unhelpful. Indeed, these are areas that have seen the most advances of late. Chapter 8 addresses hot games, games such as GO and AMAZONS in which there is a great incentive to move first, while Chapter 9 addresses *all-small* games, where the value of a move is more subtle. Chapter 10, entitled "Trimming Game Trees," describes two more recent techniques for identifying the core features of games, *reduced canonical form* and *ordinal sums*.

Chapter ω is a brief listing of other areas of active research that we could not fit into an introductory text.

In Appendix A, we present top-down induction, an approach that we use often in the text. While the student need not read the appendix in its entirety, the first few sections will help ground the format and foundation of the inductive proofs found in the text.

Appendix B is a brief introduction to CGSuite, a powerful programming toolkit written by Aaron Siegel in Java for performing algebraic manipulations on games. CGSuite is to the combinatorial game theorist what Maple or Mathematica is to a mathematician or physicist. While the reader need not use CGSuite while working through the text, the program does help to build intuition, double-check work done by hand, develop hypotheses, and handle some of the drudgery of rote calculations.

Appendix D contains the rules to many games, ordered alphabetically. In particular we include any game that appears multiple times in the text, or is found in the literature. We do not always state the detailed rules of a game in the text, so the reader will want to refer to this appendix often.

The supporting website for the book is located at www.lessonsinplay.com. Look there for links, programs, and addenda, as well as instructions for accessing the online solutions manual for instructors.

Acknowledgments

While we are listed as the *authors* of this text, we do not claim to be the main contributors. The textbook emerged from a mathematically rich environment created by others. We got to choose the words and consequently, despite the best efforts of friends and colleagues, all the errors are ours.

Many of the contributors to this environment are cited within the book. There were many who also contributed to and improved the contents of the text itself and who deserve special thanks. We are especially grateful to Elwyn Berlekamp, John Conway, and Richard Guy who encouraged — and, at times, hounded — us to complete the text, and we hope it helps spawn a new generation of active aficionados.

Naturally, much of the core material and development is a reframing of material in *Winning Ways* and *On Number and Games*. We have adopted some of the proofs of J P Grossman, particularly that of the *Number-Avoidance Theorem*. Aviezri Fraenkel contributed the *Fundamental Theorem of Combinatorial Games,* which makes its appearance at the start of Chapter 2. Dean Hickerson helped us to prove Theorem 6.15 on page 147, that a game with negative incentives must be a number. John Conway repeatedly encouraged us to adopt the *uptimal* notation in Chapter 9, and it took us some time to see the wisdom of his suggestions. Elwyn Berlekamp and David Molnar contributed some fine problems. Paul Ottaway, Angela Siegel, Meghan Allen, Fraser Stewart, and Neil McKay were students who pretested portions of the book and provided useful feedback, corrections, and clarifications. Elwyn Berlekamp,

Richard Guy, Aviezri Fraenkel, and Aaron Siegel edited various chapters of
our work for technical content, while Christine Aikenhead edited for language.
Brett Stevens and Chris Lewis read and commented on parts of the book. Susan
Hirshberg contributed the title of our book.

We also thank all those who identified typos and errors in our first edition,
especially Matthew Ferland, Ted Hwa, Ishihara Toru, and Mike Fisher.

In this age of large international publishers, A K Peters was a fantastic and
refreshing publishing house to work with on the first edition. They cared more
about the dissemination of fine works than about the bottom line. We will
miss them. But, we have been delighted to work with CRC Press and Taylor &
Francis on this edition, who have been tremendously helpful, especially given
the authors' predilection to procrastination.

The authors would like to thank their spice[2] for their loving support, and
Lila and Tovia, who are the real *Lessons in Play*.

---

[2]As in *spice of our life* — more affectionate than partner, spouse, or significant other.

Preparation for Chapter 0

Before each chapter are several quick prep problems that are worth tackling in preparation for reading the chapter.

Prep Problem 0.1. Make a list of all the two-player games you know of and classify each one according to whether or not it uses elements of chance (e.g., dice, coin flips, randomly dealt cards) and whether or not there is hidden information.

Prep Problem 0.2. Locate the textbook website, www.lessonsinplay. com, and determine whether it might be of use to you.

To the instructor: Before each chapter, we will include a few suggestions to the instructor. Usually these will be examples that do not appear in the book, but that may be worth covering in lecture. The student unsatisfied by the text may be equally interested in seeking out these examples.

We highly recommend that the instructor and the student read Appendix A on top-down induction. We present induction in a way that will be unfamiliar to most, but that leads to more natural proofs, particularly those found in combinatorial game theory.

The textbook website, www.lessonsinplay.com, has directions for how instructors can obtain a solution manual.

Chapter 0

Combinatorial Games

We don't stop playing because we grow old;
we grow old because we stop playing.

George Bernard Shaw

This book is all about *combinatorial games* and the mathematical techniques that can be used to analyze them. One of the reasons for thinking about games is so that you can play them more skilfully and with greater enjoyment; so let's begin with an example called DOMINEERING. To play you will need a chessboard and a set of dominoes. The domino pieces should be big enough to cover or partially cover two squares of the chessboard but no more. You can make do with a chessboard and some slips of paper of the right size or even play with pen or pencil on graph paper (but the problem there is that it will be hard to undo moves when you make a mistake!). The rules of DOMINEERING are simple. Two players alternately place dominoes on the chessboard. A domino can only be placed so that it covers two adjacent squares. One player, Louise, places her dominoes so that they cover vertically adjacent squares. The other player, Richard, places his dominoes so that they cover horizontally adjacent squares. The game ends when one of the players is unable to place a domino, and that player then loses. Here is a sample game on a 4×6 board with Louise moving first:

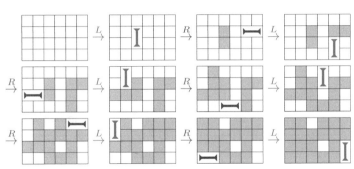

Since Louise placed the last domino, she has won.

Exercise 0.1. Stop reading! Find a friend and play some games of DOMINEER-ING. A game on a full chessboard can last a while so you might want to play on a 6×6 square to start with.

If you did the exercise, then you probably made some observations and learned a few tactical tricks in DOMINEERING. One observation is that after a number of dominoes have been placed the board *falls apart* into disconnected regions of empty squares. When you make a move you need to decide what region to play in and how. Suppose that you are the vertical player and that there are two regions of the form:

Obviously you could move in either region. However, if you move in the hook-shaped region, then your opponent will move in the square. You will have no more moves left so you will lose. If instead you move in the square, then your opponent's only remaining move is in the hook. Now you still have a move in the square to make, and so your opponent will lose. If you are L and your opponent is R, play should proceed as

This is also why an opening move such as

is good since it *reserves* the two squares in the upper left for you later. In fact, if you play seriously for a while it is quite possible that the board after the first four moves will look something like

Simply put, the aim of combinatorial game theory is to understand in a more detailed way the principles underlying the sort of observations that we have just made about DOMINEERING. We will learn about games in general and how to understand them but, as a bonus, how to play them well!

0.1 Basic Terminology

In this section we will provide an informal introduction to some of the basic
concepts and terminology that will be used in this book and a description of
how combinatorial games differ from some other types of games.

Combinatorial games

In a *combinatorial game* there are two players who take turns moving alternately. Play continues until the player whose turn it is to move has no legal
moves available. No chance devices such as dice, spinners, or card deals are
involved, and each player is aware of all the details of the game position (or
game state) at all times. The rules of each game we study will ensure that it
must end after a finite sequence of moves, and the winner is often determined
on the basis of who made the last move. In *normal play* the last player to move
wins. In *misère game play* the last player loses.

In fact, combinatorial game theory can be used to analyze some games that
do not quite fit the above description. For instance, in DOTS & BOXES, players
may make two (or more) moves in a row. Most CHECKERS positions are *loopy*
and can lead to infinitely long sequences of moves. In GO and CHESS the last
mover does not determine the winner. Nonetheless, combinatorial game theory
has been applied to analyze positions in each of these games.

By contrast, the classical mathematical theory of games is concerned with
economic games. In such games the players often play simultaneously and
the outcome is determined by a payoff matrix. Each player's objective is to
guarantee the best possible payoff against any strategy of the opponent. For a
taste of economic game theory, see Problem 5.

The challenge in analyzing economic games stems from simultaneous deci-
sions: each player must decide on a move without knowing the move choice(s)
of her opponent(s). The challenge of combinatorial games stems from the sheer
quantity of possible move sequences available from a given position.

Combinatorial game theory is most straightforward when we restrict our
attention to *short games*. In the play of a short game, a position may never
be repeated, and there are only a finite number of other positions that can be
reached. We implicitly (and sometimes explicitly) assume all games are short
in this text.

Introducing the players

The two players of a combinatorial game are traditionally called *Left* (or just *L*)
and *Right* (*R*). Various conventional rules will help you to recognize who is
playing, even without a program:

Left	Right
Louise	Richard
Positive	Negative
bLack	White
bLue	Red
Vertical	Horizontal
Female	Male
Green	
Gray	

Alice and *Bob* will also make an appearance when the first player is important (Alice moves first). To help remember all these conventions, note that despite the fact that they were introduced as long ago as the early 1980s in *Winning Ways* (*WW*) [BCG01], the chosen dichotomies reflect a relatively modern "politically correct" viewpoint.

Often we will need a neutral color, particularly in pen and paper games or games involving pieces. If the game is between blue and red then this neutral color is green (because green is good for everyone!), while if it is between black and white then the neutral color is gray (because gray is neither black nor white!). Because this book is printed in color, games traditionally played in black and white (and gray) are presented in color instead. That is,

$$\text{black} = \text{blue},$$
$$\text{white} = \text{red},$$
$$\text{gray} = \text{green}.$$

Options

If a position in a combinatorial game is given and it happens to be Left's turn to move, she will have the opportunity to choose from a certain set of moves determined by the rules of the game. For instance in DOMINEERING, where Left plays the vertical dominoes, she may place such a domino on any pair of vertically adjacent empty squares. The positions that arise from exercising these choices are called the *left options* of the original position. Similarly, the *right options* of a position are those that can arise after a move made by Right. The *options* of a position are simply the elements of the union of these two sets.

We can draw a *game tree* of a position (as a directed tree) by the following procedure.

- Create a node for the original position, and draw nodes for each of its options, placing them below the first node. Then draw a directed edge from the top node to its options.

- For each option, again draw nodes for each of *its* options, placing them below and drawing a directed edge to these new nodes.

- Repeat with any subposition that still has unexpanded options.

The nodes of the game tree correspond to the *followers* of the original position and are *all* the positions that result from any sequence of moves. The followers include the original position (the empty sequence of moves!) and also sequences in which players may get several moves in row. It is also possible to have a position appear in many places of the tree. The *game graph* is obtained by merging all the nodes that correspond to a single position, but the game tree is more important for induction purposes.

As a visual aid, our game trees will have the left options appearing below and to the left of the game and right options below and to the right. Often, induction will be based on the options, and we will draw a *partial game tree* consisting only of the original position and its options:

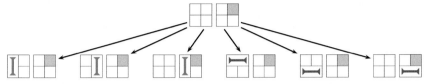

Occasionally, we will include some other interesting followers:

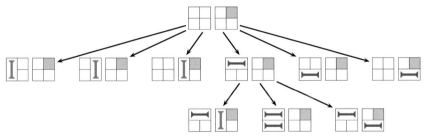

It may seem odd that we are showing two consecutive right moves in a game tree, but much of the theory of combinatorial games is based on analyzing situations where games *decompose* into several subgames. It may well be the case that in some of the subgames of such a decomposition, the players do not alternate moves.

We saw this already in the DOMINEERING "square and hook" example. Left, if she wants to win, winds up making two moves in a row in the square:

Thus, we show the game tree for a square with Left and/or Right moving twice in a row:

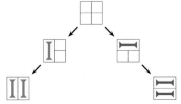

As we will see later in Chapter 4, *dominated options* are often omitted from the game tree, when an option shown is at least as good:

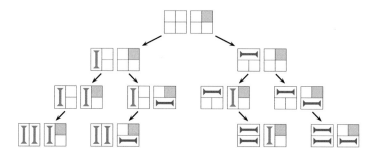

In some games the left options and the right options of a position are always the same. Such games are called *impartial*. The study of impartial combinatorial games is the oldest part of combinatorial game theory and dates back to the early twentieth century. On the other hand, the more general study of non-impartial games was pioneered by John Conway in *On Numbers and Games* (*ONAG*) [Con01] and by Elwyn Berlekamp, John Conway, and Richard Guy in *WW* [BCG01]. Since "non-impartial" hardly trips off the tongue, and "partial" has a rather ambiguous interpretation, it has become commonplace to refer to non-impartial games as *partizan games*.

To illustrate the difference between these concepts, consider a variation of DOMINEERING called CRAM. CRAM is just like DOMINEERING except that each player can play a domino in either orientation. Thus, it becomes an impartial game since there is now no distinction between legal moves for one player and legal moves for the other.

Let's look at a position in which there are only four remaining vacant squares in the shape of an L:

In CRAM the next player to play can force a win by playing a vertical domino at the bottom of the vertical strip, leaving

which contains only two non-adjacent empty squares and hence allows no further moves. In DOMINEERING if Left (playing vertically) is the next player, she can win in exactly this way. However, if Right is the next player his only legal move is to cover the two horizontally adjacent squares, which still leaves a move available to Left. So (assuming solid play) Left will win regardless of who plays first:

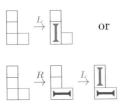 or

Much of the theory that we will discuss is devoted to finding methods to determine who will win a combinatorial game assuming sensible play by both sides. In fact, the eventual loser has no really *sensible* play[1] so a *winning strategy* in a combinatorial game is one that will guarantee a win for the player employing it no matter how his or her opponent chooses to play. Of course, such a strategy is allowed to take into account the choices actually made by the opponent — to demand a uniform strategy would be far too restrictive!

Problems

1. Consider the position

   (a) Draw the complete game trees for both CRAM and DOMINEERING. The leaves (bottoms) of the tree should all be positions in which neither player can move. If two left (or right) options are symmetrically identical, you may omit one.

   (b) In the position above, who wins at DOMINEERING if Vertical plays first? Who wins if Horizontal plays first? Who wins at CRAM?

2. Suppose that you play DOMINEERING (or CRAM) on *two* 8 × 8 chessboards. At your turn you can move on either chessboard (but not both!). Show that the second player can win.

3. Take the ace through five of one suit from a deck of cards and place them face up on the table. Play a game with these as follows. Players alternately pick a card and add it to the right-hand end of a row. If the row ever contains a sequence of three cards in increasing order of rank (ace is low), or in decreasing order of rank, then the game ends and the player who formed that sequence is the winner. Note that the sequence

---

[1] Unless he has some ulterior motive not directly related to the game such as trying to make it last as long as possible so that the bar closes before he has to buy the next round of drinks.

need not be consecutive either in position or value, so, for instance, if the play goes $4, 5, 2, 1$ then the $4, 2, 1$ is a decreasing sequence.

    (a) Show that this is a proper combinatorial game (the main issue is to show that draws are impossible).

    (b) Show that the first player can always win.

4. Start with a heap of counters. As a move from a heap of n counters, you may either:

    • assuming n is not a power of 2, remove the largest power of 2 less than n; or

    • assuming n is even, remove half the counters.

Under normal play, who wins? How about misère play?

5. The goal of this problem is to give the reader a taste of what is *not* covered in this book. Two players play a 2×2 *zero-sum matrix game*. (*Zero sum* means that whatever one person loses, the other gains.) The players are shown a 2×2 matrix of positive numbers. Player A chooses a row of the matrix, and player B simultaneously chooses a column. Their choice determines one matrix entry, that being the number of dollars B must pay A. For example, suppose the matrix is

$$\begin{pmatrix} 1 & 4 \\ 3 & 2 \end{pmatrix}.$$

If player A chooses the first row with probability $\frac{1}{4}$, then no matter what player B's strategy is, player A is *guaranteed* to get an average of $2.50. If, on the other hand, player B chooses the columns with 50-50 odds, then no matter what player A does, player B is *guaranteed* to have to pay an average of $2.50. Further, neither player can guarantee a better outcome, and so B should pay player A the fair price of $2.50 to play this game.

In general, if the entries of the matrix game are

$$\begin{pmatrix} a & b \\ c & d \end{pmatrix},$$

as a function of a, b, c, and d, what is the fair price that B should pay A to play? (Your answer will have several cases.)

Preparation for Chapter 1

Prep Problem 1.1. Play DOTS & BOXES with a friend or classmate. The rules are found on page 307 of Appendix D. You should start with a 5 × 6 grid of dots. You should end up with a 4 × 5 grid of 20 boxes, so the game might end in a tie.

When playing a game for the first time, feel free to move quickly to familiarize yourself with the rules and to get a sense for what can happen in the game.

After a few games of DOTS & BOXES, write a few sentences describing any observations you have made about the game. Perhaps you found a juncture in the game when the nature of play changes? Did you have a strategy? (It need not be a good strategy.)

Prep Problem 1.2. Play SNORT with a friend or classmate. The rules are found on page 313 of Appendix D. (Note that if the Winner is not specified in a ruleset, you should assume normal play, that the last legal move wins.) You should play on paths of various lengths: for instance,

Jot down any observations you have about the game, and then try playing COL on the same initial positions.

Prep Problem 1.3. Play CLOBBER with a friend or classmate. The rules are found on page 305 of Appendix D. You should start with a 5 × 6 grid of boxes:

Jot down any observation you have about the game.

Prep Problem 1.4. Play NIM (rules on page 311) with a friend or classmate. Begin with the three heap position with heaps of sizes 3, 5, and 7.

To the instructor: While DOTS & BOXES is a popular topic among students, it also takes quite a bit of time to appreciate. View the topic as optional. If you do cover it, allow time for students to play practice games. Another option is to cover it later in the term before a holiday break.

Chapter 1

Basic Techniques

If an enemy is annoying you by playing well,
consider adopting his strategy.

Chinese proverb

There are some players who seem to be able to play a game well immediately
after learning the rules. Such gamesters have a number of tricks up their sleeves
that work well in many games without much need for further analysis. In this
chapter we will teach you some of these tricks or, to use a less emotive word,
heuristics.

Of course, the most interesting games are those to which none of the heuristics apply directly, but knowing them is still an important part of getting started
with the analysis of more complex games. Often, you will have the opportunity
to consider moves that lead to simple positions in which one or more of the
heuristics apply. Those positions are then easily understood, and the moves
can accordingly be taken or discarded.

1.1   Greedy

The simplest of the heuristic rules or strategies is called the *greedy strategy*.
A player who is playing a greedy strategy grabs as much as possible whenever
possible. Games that can be won by playing greedily are not terribly interesting
at all — but most games have some aspects of greedy play in them.   For
instance, in CHESS it is almost always correct to capture your opponent's queen
with a piece of lesser value (taking a greedy view of "getting as much material
advantage as possible"), but not if doing so allows your opponent to capture
your queen, or extra material, or especially not if it sets up a checkmate for the
opponent. Similarly, the basic strategy for drawing in TIC TAC TOE is a greedy

one based on the idea "always threaten to make at least one line, or block any threat of your opponent."

Definition 1.1. A player following a *greedy strategy* always chooses the move that maximizes or minimizes some quantity related to the game position after the move has been made.

Naturally, the quantity on which a greedy strategy is based should be easy enough to calculate that it does not take too long to figure out a move. If players accumulate a *score* as they play (where the winner is the one who finishes with the higher score), then that score is a natural quantity to try to maximize at each turn.

Does a greedy strategy always work? Of course not, or you wouldn't have a book in front of you to read. But in some very simple games it does. In the game GRAB THE SMARTIES[1] each player can take at his or her move any number of Smarties from the box, provided that they are all the same color. Assuming that each player wants to collect as many Smarties as possible, the greedy strategy is ideal for this sort of game. You just grab all the Smarties of some color, and the color you choose is the one for which your grab will be biggest.

Sometimes though, a little subtlety goes a long way.

Example 1.2. Below is the board after the first moves in a very boring game of DOTS & BOXES:

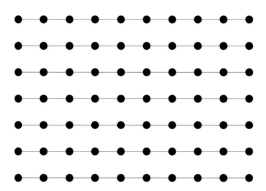

Suppose that it is now Alice's turn. No matter where Alice moves, Bob can take all the squares in that row. If he does so, he then has to move in another row, and Alice can take all the squares in this row. They trade off in this way until they both have 27 boxes and the game is tied:

---

[1]An American player might play the less tasty variant, GRAB THE M&M'S.

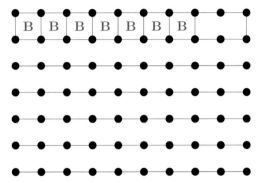

But Bob is being too greedy. Instead of taking all the squares in the row that Alice opens up, he should take all but two of them, then make the *double-dealing move*, which gives away the last two boxes to Alice. For example, if Alice moves somewhere toward the left-hand end of the first row, Bob replies with

Alice now has a problem. Regardless of whether she takes the two boxes that Bob has left for her, she still has to move first in another row. So she might as well take the last two in a *double-cross* (since otherwise Bob will get them on his next turn), but she then has to give seven boxes of some other row to Bob in his next turn. By repeating this strategy for each but the last row, where he takes all the boxes, Bob finishes with $5 \times 7 + 9 = 44$ boxes while Alice gets only $5 \times 2 = 10$ boxes:

Again, a move that makes two boxes with one stroke is called a *double-cross*. (A person who makes such a play might feel double-crossed for having to reply to a *double-dealing move*.)

It is not a bad idea to use a greedy strategy on your first attempt at playing a new game, particularly against an expert. It is easy to play, so you won't waste time trying to figure out good moves when you don't know enough about the game to do so. And when the expert makes moves that refute your strategy (i.e., exposes the traps hidden in the greedy strategy), then you can begin to understand the subtleties of the game.

Exercise 1.3. Now that you have learned a bit of strategy, play DOTS & BOXES against a friend. (If you did the prep problems for this chapter, you already played once. Now, you may be able to play better.)

1.2   Symmetry

A famous CHESS wager goes as follows: An unknown CHESS player, Jane Pawn-pusher, offers to play two games, playing the white pieces against Garry Kasparov and black against Anatoly Karpov simultaneously. She wagers $1 million dollars that she can win or draw against one of them. Curiously, she can win the wager without knowing much about CHESS. How?

What she does is simply wait for Karpov to make a move (white moves first in CHESS), and whatever Karpov does, she makes the same move as her first move against Kasparov. Once Kasparov replies, she plays Kasparov's reply against Karpov. If Kasparov beats her, she will beat Karpov the same way.[2] A strategy that maintains a simple symmetry like this is called *Tweedledum-Tweedledee* or *copycat*.

Example 1.4. The Tweedledum-Tweedledee strategy is effective in two-heap NIM. If the two heaps are of the same size, then you should invite your opponent to move first. She must choose a heap and remove some counters. You choose the other heap and take away the same number of counters leaving two equal sized heaps again. On the other hand, if the game begins with two heaps of different sizes, you should rush to make the first move, taking just enough counters from the larger heap to make them equal. Thereafter, you adopt the Tweedledum-Tweedledee approach.

Symmetry is an intuitively obvious strategy. Whenever your opponent does something on one part of the board, you should mimic this move in another part. Deciding how this mimicry should happen is the key. To be played

---

[2]Readers familiar with cryptography may observe Jane is making a man-in-the-middle attack.

successfully, you should not leave a move open to your opponent that allows him to eliminate your mimicking move.

Example 1.5. If Blue moves from the 3×4 CLOBBER game

to any of the three positions

then Red can play the remainder of the game using a 180 degree symmetry strategy. This establishes that each of these three moves for Blue was a poor choice on her first turn. In fact, from this position it happens to be the case that she simply has no good first moves, but the rest of her initial moves cannot be ruled out so easily due to symmetry.

Sometimes, symmetry can exist that is not apparent in the raw description of a game.

Example 1.6. Two players take turns putting checkers down on a checkerboard. One player plays blue, one plays red. A player who completes a 2×2 square with four checkers of one color wins.

This game should end in a draw. First, imagine that most of the checkerboard is tiled with dominoes using a brickwork pattern:

If your opponent plays a checker in a domino, you respond in the same domino. If you cannot (because you move first, or the domino is already filled, or your opponent fails to play in a domino), play randomly. Since every 2×2 square contains one complete domino, your opponent cannot win. Therefore, both players can force at least a draw, and neither one can force a win.

Exercise 1.7. Two players play $m \times n$ CRAM.

(a) If m and n are even, who should win? The first player or the second player? Explain your answer.

(b) If m is even and n is odd, who should win? Explain.

(When m and n are odd, the game remains interesting.)

1.3 Parity

Parity is a critical concept in understanding and analyzing combinatorial games. A number's *parity* is whether the number is odd or even. In lots of games, only the parity of a certain quantity is relevant — the trick is to figure out just what quantity! With the normal play convention that the last player with a legal move wins, it is always the objective of the first player to play to ensure that the game lasts an odd number of moves, while the original second player is trying to ensure that it lasts an even number of moves.

This is part of the reason why symmetry as we mentioned earlier is also important — it allows the second player (typically) to think of moves as being blocked out in pairs, ensuring that he has a response to any move his opponent might make.

The simplest game for which parity is important is called SHE LOVES ME SHE LOVES ME NOT. This game is played with a single daisy. The players alternately remove exactly one petal from the daisy and the last player to remove a petal wins. Obviously, all that matters is the original parity of the number of petals on the daisy. If it is odd then the first player will win; if it is even then the second player will win.

More usually SHE LOVES ME SHE LOVES ME NOT is delivered in some sort of disguise.

Example 1.8. Take a heap of 29 counters. A move is to choose a heap (at the start there is only one) and split it into two non-empty heaps. Who wins?

Imagine the counters arranged in a line. A move effectively is to put a bar between two counters. This corresponds to splitting a heap into two: the counters to the left and those to the right up to the next bar or end of the row. There are exactly 28 moves in the game. The game played with a heap of n counters has exactly $n - 1$ moves! The winner is the first player if n is even and the second player if n is odd.

Exercise 1.9. A chocolate bar is scored into smaller squares or rectangles. (Lindt's Swiss Classic, for example, is 5×6.) Players take turns picking up one piece (initially the whole bar), breaking the piece into two along a scored line, and setting the pieces back down. The player who moves last wins. Our goal is to determine *all* winning moves.

1.4 Give Them Enough Rope!

The previous strategies are all explicit, and when they work, you can win the game. This section is about confounding your opponent in order to gain time for analysis.

If you are in a losing position, it pays to follow the *Enough Rope Principle:* Make the position as complicated as you can with your next move.[3] Hopefully, your opponent will tie himself up in knots while trying to analyze the situation.

For example, suppose you are Blue and are about to move from the following CLOBBER position:

If you are more astute than the authors, you could conclude that you have no winning moves. However, you should probably not throw in the towel just yet. But you also should not make any moves for which your opponent has a simple strategy for winning. If your opponent has read this chapter, you should avoid capturing an edge piece with your center piece, for then Red can play a rotational symmetry strategy. However, there are several losing responses from either of the positions

 or

and so these moves, while losing, are reasonable.

The Enough Rope Principle has other implications as well. If you are confused about how best to play, do not simplify the position to the point where your opponent will not be confused, especially if you are the better player.

The converse applies as well. If you are winning, play moves that simplify. Do not give you opponent opportunities to complicate the position, lest you be hoist by your own petard.

Don't give them any rope

This is contrary to the advice in the rest of the section. If you do not know that you are losing the game and you are playing against someone of equal or less game-playing ability, then a very good strategy is to move so as to restrict the number of options that your opponent has and increase the number of your own options. This is a heuristic that is often employed in first attempts to produce programs that will play games reasonably well. This has been used in AMAZONS, CONNECT-4, and OTHELLO.

1.5 Strategy Stealing

Strategy stealing is a technique whereby one player steals another player's strategy. Why would you want to steal a strategy? Let's see . . . .

---

[3]At least one of the authors feels compelled to add, *except if you are playing against a small child.*

Two-move chess

Players play by the ordinary rules of chess, but each player plays two consecutive regular chess moves on each turn (this example appears in [SA03]).

White, who moves first, can under perfect play win or draw. For if Black had a winning strategy, White could steal it by playing a Knight out and then back again. Black is now faced with making the initial foray on the board with the roles of Black and White reversed.

Chomp

The usual starting position of a game of CHOMP consists of a rectangle with one poison square in the lower-left corner:

A move in CHOMP is to choose a square and to remove it and all other squares above or to the right of it. A game between players Alice and Bob might progress as follows:

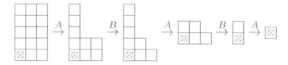

And Bob loses for he must take the poison square.

Theorem 1.10. CHOMP, *when played on a rectangular board larger than* 1×1, *is a win for the first player.*

Proof: Suppose that the first player chomps *only* the upper-right square of the board. If this move wins, then it is a first-player win. If, on the other hand, this move loses, then the second player has a winning response of chomping all squares above or to the right of some square, x. But move x was available to the first player on move one, and it removes the upper-right square, so the first player has move x as a winning first move.   □

This is a *non-constructive proof* in that the proof gives no information about what the winning move is. The proof can be rephrased as a *guru argument,* echoing the CHESS wager of Section 1.2.

Bridg-it

The game of BRIDG-IT is played on two offset grids of blue and red dots. Here is a position after five moves (Blue played first):

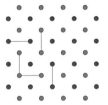

The players, Blue and Red, alternate drawing horizontal or vertical lines joining adjacent dots of the player's chosen color. Blue wins by connecting any dot in the top row to any dot in the bottom row by a path. Red is trying to connect the left side to the right side. In the following position, Blue has won, with a path near the right side of the board:

Lemma 1.11. BRIDG-IT *cannot end in a draw.*

Sketch of proof: The game is unaffected if we consider the top and bottom rows of blue dots as connected. Suppose the game has ended, and neither player has won. Let S be the set of nodes that Red can reach from the left side of the board. Then, starting from the upper-left blue dot, Blue can go from the top to the bottom edge by following the boundary of the set S. As an example, set S consists of the red dots below, and the blue path following the boundary is shown on the right:

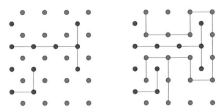

All the edges connecting blue dots must be present, for otherwise set S could be extended. □

Theorem 1.12. *The first player wins at* BRIDG-IT, *where the starting board is an $n \times (n+1)$ grid of red dots overlapping an $(n+1) \times n$ grid of blue dots.*

Proof: Note that the board is symmetric when reflected about the main diagonal. If the second player has a winning strategy, the first player can adopt it. In particular, before her first move, the first player pretends that some random invisible move x has been made by the opponent and then responds as the second player would have. If the opponent's n^{th} move is move x, then the first player pretends that the opponent actually played the nth move at some other location x'. Continuing in this fashion, the first player has stolen the winning second-player strategy, with the only difference that the opponent always has one fewer lines on the board. This missing move can be no worse for the first player, and so she will win.   □

There are explicit winning strategies for BRIDG-IT, avoiding the need for a non-constructive strategy-stealing argument. So, not only can you find out you *should* win, but by utilizing a bit of elementary graph theory, you can also find out *how* to win! See, for example, [BCG01, volume 3, pp. 744–746] or [Wes01, pp. 73–74].

1.6   Change the Game!

Sometimes a game is just another game in disguise. In that case one view can be more, or less, intuitive than the other.

Example 1.13. In 3-TO-15, there are nine cards, face up, labeled with the digits $\{1, 2, 3, \ldots, 9\}$. Players take turns selecting one card from the remaining cards. The first player who has three cards adding up to 15 wins.

This game should end in a draw. Surprisingly, this is simply TIC TAC TOE in disguise! To see this, construct a *magic square* where each row, column, and diagonal add up to 15:

4	9	2
3	5	7
8	1	6

You can confirm that three numbers add up to 15 if and only if they are in the same TIC TAC TOE line. Thus, you can treat a play of 3-TO-15 as play of TIC TAC TOE. Suppose that you are moving first. Choose your TIC TAC TOE move, note the number on the corresponding square, and select the corresponding card. When your opponent replies by choosing another card, mark the TIC TAC TOE board appropriately, choose your TIC TAC TOE response, and again take the corresponding card. Proceed in this fashion until the game is over. So if you can play TIC TAC TOE, you can play 3-TO-15 just as well.

Exercise 1.14. Play 3-TO-15 against a friend. As you and your friend move, mark the magic square TIC TAC TOE board with Xs and Os to convince yourself that the games really are the same.

Example 1.15. COUNTERS is played with tokens on a $1 \times n$ strip of squares. Players can put down a counter on an empty square or move a counter leftward to the next empty square. In the latter case all the counters that it jumps over (if any) are removed. POWER 2 is a game whose position is a positive integer n and a move is to subtract a power of 2 from n provided that the result is non-negative.

These games are the same! For example, consider the position 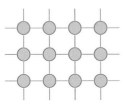. Change empty spaces to 1's and counters to 0's and the position is 01001, which is the binary expansion for 9 (with a leading zero). The possible moves (and their corresponding numbers) are

$$
\begin{aligned}
&= 00001 = 1 = 9 - 8,\\
&= 00101 = 5 = 9 - 4,\\
&= 00111 = 7 = 9 - 2, \text{ and}\\
&= 01000 = 8 = 9 - 1.
\end{aligned}
$$

Problem 10 of Chapter 7 asks you to solve POWER 2.

1.7   Case Study: Long Chains in Dots & Boxes

We already observed in Example 1.2 on page 12 that the first player to play on a long chain in a DOTS & BOXES game typically loses. In this section, we will investigate how that can help a player win against any first grader.

First, consider a dual form of DOTS & BOXES called STRINGS & COINS. Here is a typical starting position:

A move in STRINGS & COINS consists of cutting a string. If a player severs the last of the four strings attached to a coin, the player gets to pocket the coin and must move again. This game is the same as DOTS & BOXES but disguised: a coin is a box, and cutting a string corresponds to drawing a line between two boxes. For example, here is a DOTS & BOXES position and its dual STRINGS & COINS position:

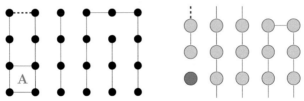

Drawing the dotted line in DOTS & BOXES corresponds to cutting the dotted string in STRINGS & COINS. We will investigate STRINGS & COINS positions, since people tend to find that important properties of the positions (such as long chains) are easier to visualize and identify in this game than in DOTS & BOXES.

Positions of the following form are termed *loony:*

○─○─┆ANY┆ is loony except ○─○─○ is not loony.

The hidden portion of the position (in the box marked ANY) can be any position except a single coin. The defining characteristic shared by all loony positions is that the next player to move has a choice of whether or not to make a double-dealing move. A *loony move,* denoted by a 𝔇, is any move to a loony position. All loony moves are labeled in the following DOTS & BOXES and equivalent STRINGS & COINS positions:

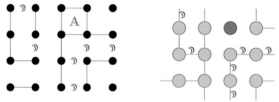

To summarize, if Bob makes a loony move (a move to a loony position), Alice may (or may not) reply with a double-dealing move. From there, Bob might as well double-cross before moving elsewhere.

Loony move	Double-dealing move	Double-cross

(Bob must move again after the double-cross.)

Theorem 1.16. *Under optimal play from a loony position, the player to move next can get at least half the remaining coins.*

Proof: Suppose that player Alice is to move next from the loony position

$$\bigcirc\!\!-\!\!\bigcirc\!\!-\!\!\boxed{\text{ANY}}$$

Consider playing only on the hidden position in the box marked ANY (without the two extra coins). Suppose that the player moving next from ANY can guarantee pocketing n coins. In the full position, Alice has at least two choices:

- Pocket the two coins (making two cuts) and move on ANY, pocketing n more coins for a total of $n + 2$:

$$\bullet\qquad\bullet\qquad\boxed{\text{ANY}}$$

- Sacrifice the two coins, cutting off the pair in one move. Whether or not the opponent chooses to pick up the two coins, he must move next on ANY, and so the most he can pocket is $n + 2$ coins:

$$\bigcirc\!\!-\!\!\bigcirc\qquad\boxed{\text{ANY}}$$

Thus, Alice can collect $n + 2$ coins, or all but $n + 2$ coins. One of these two numbers is at least half the total number of coins! $\square$

In practice, the player about to move often wins decisively, especially if there are very long chains.

Hence, from most positions, a 𝔇 move (i.e., a move to a 𝔇 position) is a losing move and might as well be illegal when making a first pass at understanding a position.

A *long chain* consists of $k \geq 3$ coins and *exactly $k + 1$ strings* connected in a line:

$$-\!\bigcirc\!\!-\!\!\bigcirc\!\!-\!\!\bigcirc\!\!-\cdots-\!\bigcirc\!\!-$$

Notice that any move on a long chain is 𝔇.

Exercise 1.17. Find the non-𝔇 move(s) on a (short) chain of length 2:

Exercise 1.18. Here are two separate STRINGS & COINS positions. Alice is about to play in each game:

(a) Both positions are loony. Explain why.

(b) In one of the positions, Alice should make a double-dealing move. Which one? Why?

(c) Estimate the score of a well-played game from each of the two positions. (Alice should be able to win either game.)

So, it is crucial to know whose move it is if only long chains remain, for in such a position all moves are loony and Theorem 1.16 tells us that the player about to move will likely lose. To this end, consider a position with only long chains. We distinguish a move (drawing a line or cutting a string) from a turn, which may consist of several moves.

Define

$$
\begin{aligned}
M^- &= \text{number of moves played so far;} \\
M^+ &= \text{number of moves remaining to be played;} \\
M &= M^+ + M^- = \text{possible moves from the start position;} \\
T &= \text{number of turn transfers so far;} \\
B^- &= \text{number of boxes (or coins) taken already;} \\
B^+ &= \text{number of boxes (or coins) left to be taken;} \\
B &= B^+ + B^- = \text{total number of boxes (or coins) in the start position;} \\
C &= \text{number of long chains;} \\
D &= \text{number of double-crosses so far.}
\end{aligned}
$$

Recall that double-crosses are single moves that take two coins in one cut (or complete two boxes in one stroke):

We can compute the above quantities for the following position:

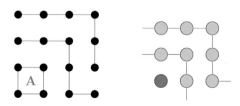

$$
\begin{aligned}
M^- &= 14, \\
M^+ &= 10, \\
M &= 24, \\
T &= 13, \\
B^- &= 1, \\
B^+ &= 8, \\
B &= 9, \\
C &= 2, \\
D &= 0.
\end{aligned}
$$

(If there were two adjacent boxes taken by the same player, we would not know if $D = 0$ or $D = 1$ without having watched the game in progress.)

We will now proceed to set up equations describing (as best we can) the state of affairs when we are down to just long chains.

- Since every long chain has one more string than coin, we know that

$$M^+ = C + B^+.$$

- Since every move either completes a turn, completes a box, or completes two boxes,

$$M^- = T + B^- - D.$$

Adding these equations, we conclude that

$$M = C + T + B - D.$$

Whose turn it is depends only on whether T is even or odd; M and B are fixed at the start of the game, so whose turn it is is determined by the number of long chains and the number of double crosses. We have all but proved the following:

Theorem 1.19. *If a* STRINGS & COINS *(or* DOTS & BOXES*) position is reduced to just long chains, player P can earn most of the remaining boxes, where*

$$P \equiv M + C + B + D \pmod 2,$$

the first player to move is player $P = 1$, and her opponent is player $P = 2$ (or, if you like, $P = 0$).

Proof: By the discussion preceding the theorem,

$$M = C + T + B - D.$$

If all that remains are long chains, whoever is on move (i.e., about to move) must make a loony move, which by Theorem 1.16 guarantees that the last player can take at least half the remaining coins. If you are player 1, say, then your opponent is on move if an odd number of turns have gone by since the start of the game; i.e., T is odd. Note that T is odd if and only if $M - C - B + D$ is odd; i.e., if and only if $P \equiv M + C + B + D$ (mod 2). (Replace odd by even for $P = 2$.)   □

In a particular game, viewed from a particular player's perspective, P, M, and B are all constant, and D is nearly always 0 (until someone makes a loony move.) So, the parity of T depends only on C, a quantity that depends on the actual moves made by the players.

In summary, when you sit down to a game of DOTS & BOXES, count

$$P + M + B,$$

where P is your player number. You seek to ensure that the parity of C, the number of long chains, matches this quantity.[4] That is, you want to arrange that $C \equiv P + M + B$ (mod 2). If you can play to make the parity of the long chains come out in your favor, you will usually win.

An example is in order. Alice played first against Bob, and they reach the following position with Alice to play:

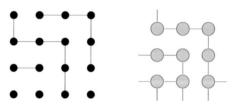

At the start of the game, Alice computed $P + M + B = 1 + 24 + 9$, an even quantity, and therefore knows that she wishes for an even number of long chains. Having identified all loony moves, she knows that the chain going around the upper and right sides will end in one long chain *unless* someone makes a loony (losing) move. So, she hopes that the lower-left portion of the board ends in a long chain. Two moves will guarantee that end, those marked below with dashed lines:

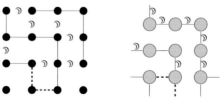

Of the two moves, Alice prefers the horizontal move, since that lengthens the chain; because she expects to win most of the long chains, longer chains favor

---

[4]In rectangular DOTS & BOXES boards, the number of dots is $1 + M + B$ (mod 2), and some players prefer to count the dots. This is the view taken in [Ber00].

her. If Bob is a beginner, a typical game might proceed as

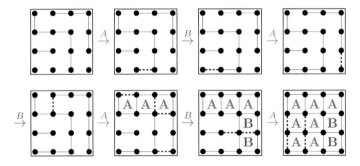

with Alice winning 7 to 2.

A more sophisticated Bob might recognize that he has lost the game of long chains and might try to finagle a win by playing loony moves earlier. This has the advantage of giving Alice fewer points for her long chains. A sample game between sophisticated players might go

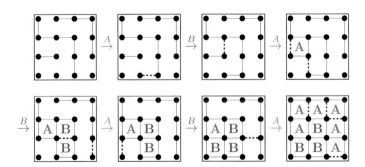

Not only does Bob lose by only 6 to 3, but Bob might win if Alice hastily takes all three boxes in the first long chain!

Exercise 1.20. What is the final score if Alice takes all three boxes instead of her first double-dealing move in the last game? Assume that both players play their best thereafter.

Suppose that Alice fails to make a proper first move. Bob can then steal control by sacrificing two boxes (without making a loony move), breaking up the second long chain. For example, play might proceed as

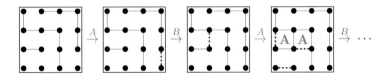

In this game, Alice should get the lower-left four boxes, but Bob will get the entire upper-right chain, winning 5 to 4.

Lastly, note that nowhere in the discussion leading up to or in the proof of Theorem 1.19 did we use any information about what the start position was. Consequently, if you come into a game already in play, you can treat the current position as the start position! In our example game between Alice and Bob repeated below, since M, the number of moves available from this start position, is fourteen and B, the number of boxes still available, is nine, the player on move (who we dub player 1 from this start position) wants an even number of long chains:

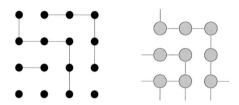

Warning: Not all games reach an endgame consisting of long chains. There are other positions in which all moves are loony. See, for example, Problem 18. These sorts of positions come up more often the larger the board size.

Enough Rope Principle revisited

In the following DOTS & BOXES (or the equivalent STRINGS & COINS) position, Bob has stumbled into Alice's trap, and Alice is now playing a symmetry strategy. Note that the upper-right box and the lower-left box are, in fact, equivalent. If this is not obvious from the DOTS & BOXES position, try looking at the corresponding STRINGS & COINS position, where the upper-right dangling string could extend toward the right without changing the position:

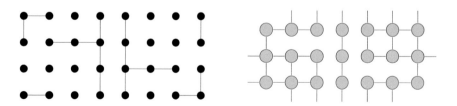

If Bob allows Alice to keep symmetry to the end of the game, Alice will succeed in getting an odd number of long chains and win. So, Bob should make a loony move now on the long chain, forcing Alice to choose between taking the whole chain or making one box and a double-dealing move:

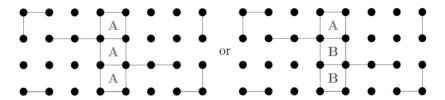

While Theorem 1.16 guarantees that Alice can win from one of the two positions (Alice first takes one box and then uses the theorem to assure half the remainder), the proof of the theorem gives no guidance about *how* to win.

If Alice chooses the first option, it is now her move on the rest of the board; she cannot play symmetry. If, on the other hand, she chooses the second option, Bob has gained a one-point advantage, which he may be able to parlay into a win.

Problems

Note that a few problems require some familiarity with graph theory. In particular, Euler's Formula, Theorem A.7 on page 263, will come in handy.

1. Consider the $2 \times n$ CLOBBER position

Show that if n is even then

is a second-player win. (By the way, the first player wins when $n \le 13$ is odd and, we conjecture, for all n odd.)

2. Prove that Left to move can win in the COL position

3. Suppose that two players play STRINGS & COINS with the additional rule that a player, on her turn, can spend a coin to end her turn. The last player to play wins. (Spending a coin means discarding a coin that she has won earlier in the game.)

(a) Prove that the first player to take any coin wins.

(b) Suppose that the players play on an m-coin by n-coin board with the usual starting position. Prove that if $m + n$ is even, the second player can guarantee a win.

(c) Prove that if $m + n$ is odd, the first player can guarantee a win.

4. Two players play the following game on a round tabletop of radius R. Players take turns placing pennies (of unit radius) on the tabletop, but no penny is allowed to touch another or to project beyond the edge of the table. The first player who cannot legally play loses. Determine who should win as a function of R.[5]

5. Who wins SNORT when played on a path of length n?

How about an $m \times n$ grid?

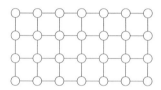

6. The game of ADD-TO-15 is the same as 3-TO-15 (page 20) except that the first player to get *any* number of cards adding to 15 wins. Under perfect play, is ADD-TO-15 a first-player win, second-player win, or draw?

7. The following vertex-deletion game is played on a directed graph. A player's turn consists of removing any single vertex with even indegree (and any edges into or out of that vertex). Determine the winner if the start position is a directed tree, with all edges pointing toward the root.

8. Two players play a vertex-deletion game on an undirected graph. A turn consists of removing exactly one vertex of even degree (and all edges incident to it). Determine the winner.

9. A bunch of coins is dangling from the ceiling. The coins are tied to one another and to the ceiling by strings as pictured below. Players alternately cut strings, and a player whose cut causes any coins to drop to the ground loses. If both players play well, who wins?

---

[5]The players are assumed to have perfect fine motor control!

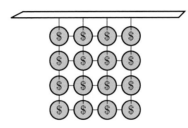

10. The game of SPROUTLETTES is played like SPROUTS, only the maximum allowed degree of a node is only 2. Who wins in SPROUTLETTES?

11. Find a winning strategy for BRUSSELS SPROUTS. (*Hint:* Describe how end positions must look and deduce how many moves the game lasts.)

12. How many moves does a game of SPROUTS last as a function of both the number of initial dots and the number of isolated degree-2 nodes at the end of the game? Give a rule for playing SPROUTS analogous to the number of long chains in DOTS & BOXES.

13. Prove that the first player wins at HEX. You are free to find and present a proof that you find in the literature, but be sure to cite your source and rephrase the argument in your own words.

14. SQUEX is a game like HEX but is played on a square board. A player makes a turn by placing a checker of her own color on the board. Squares on the board are *adjacent* if they share a side. Blue's goal is to connect the top and bottom edges with a path of blue checkers, while Red wishes to connect the left and right edges with red checkers.

    (a) Prove that the first player should win or draw an $n \times n$ SQUEX position.

    (b) For what values of n is $n \times n$ SQUEX a win for the first player, and when is it a draw? Prove your answer by giving an explicit strategy for the first player to win or the second player to draw as appropriate.

    (c) How about $m \times n$ SQUEX?

15. Alice is about to make a move in the following DOTS & BOXES position:

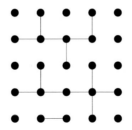

(a) Construct the equivalent STRINGS & COINS position.

(b) Determine if Alice wants an even or odd number of long chains.

(c) Determine all of Alice's winning first moves from this position.

16. You are about to make a move from the following STRINGS & COINS position:

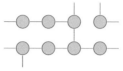

(a) Determine all winning moves.

(b) Draw the corresponding DOTS & BOXES position.

(c) How many coins/boxes should you get in a well-played game?

17. Determine all winning moves in the following DOTS & BOXES position; for your convenience, the matching STRINGS & COINS is shown on the right, which has 16 coins and 24 strings:

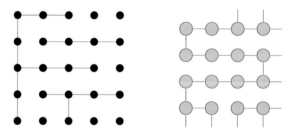

18. There are other DOTS & BOXES (or STRINGS & COINS) positions where every move is loony. For example, there can be cycles

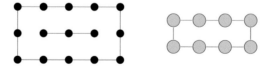

and many-legged *spiders*

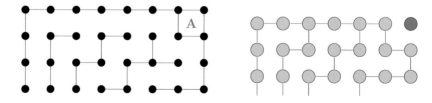

How can you adapt Theorem 1.19 and its proof to account for these positions? Note that much of the proof is found in the paragraphs preceding the theorem.

Preparation for Chapter 2

Prep Problem 2.1. Recruit an opponent and play impartial CUT-THROAT on star graphs. You can choose any start position you wish; here is one possibility:

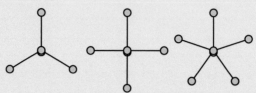

Prep Problem 2.2. Play PARTIZAN ENDNIM. You can roll a six-sided die five times to create a random five-heap position, or start from a particular position such as the six-heap position 254653.

To the instructor: In place of Section 2.6, consider covering theorems about the outcome classes of *partizan subtraction games* from [FK87]. In particular, Theorem 4 and perhaps Theorems 5 and 6 from that paper are appropriate.

Chapter 2
Outcome Classes

There's just one thing I've got to know.
Can you tell me please, who won?

Crosby, Stills, and Nash in *Wooden Ships*

In this book, we are usually concerned with who wins if both players *play perfectly*. What does playing perfectly mean? One aspect of playing perfectly is clear: if a player can force a win, then she makes a move that allows her to force a win. What if there is no such move available? In real games it is often then good play to make life as difficult as possible for your opponent, i.e., play according to what we called the Enough Rope Principle. In theory, since your opponent is assumed to be playing perfectly, such moves cannot help. So, if a player cannot force a win, then playing perfectly simply means "making a move." We formalize this discussion in the *Fundamental Theorem of Combinatorial Games* for partizan, finite, acyclic games (without draws):

Theorem 2.1. (Fundamental Theorem of Combinatorial Games) *In a game played between Albert and Bertha, with Albert moving first, either Albert can force a win moving first, or Bertha can force a win moving second, but not both.*

Proof: Each of Albert's moves is to a position which, by induction, is either a win for Bertha playing first or a win for Albert playing second. If any of his moves belong to the latter category, then by choosing one of them Albert can force a win. On the other hand, if all of his moves belong to the first category, then Bertha can force a win by using her winning strategy in the position resulting from any of Albert's moves. □

We will use this theorem implicitly many times throughout the book.

Suppose that we fix a position G, and see what happens when the first player is Left and when the first player is Right. According to the Fundamental

Theorem, there are then four possibilities, which we can use to categorize this (or any other) position into one of four *outcome classes:*

Class	Name	Definition
$\mathcal{N}$	Fuzzy	The $\mathcal{N}$ext player to play whether it be Left or Right (i.e., the 1st to play) can force a win
$\mathcal{P}$	Zero	The $\mathcal{P}$revious player who played (or 2nd to play) can force a win
$\mathcal{L}$	Positive	$\mathcal{L}$eft can force a win regardless of who moves first
$\mathcal{R}$	Negative	$\mathcal{R}$ight can force a win regardless of who moves first

At first glance, the possibility exists that not all four of these outcome classes arise in actual games. However, it is not too difficult to come up with examples of each type even in DOMINEERING:

Exercise 2.2. What is the outcome type of a two-heap NIM position with a counters in one heap and b in the other (a and b are arbitrary positive integers)?

Exercise 2.3. What are the smallest (in the sense of number of unoccupied cells) DOMINEERING positions of each outcome type?

Another way to view the different outcome classes is shown in the next table:

Outcome classes		*When Right moves first*	
		Right wins	Left wins
When Left moves first	Left wins	$\mathcal{N}$	$\mathcal{L}$
	Right wins	$\mathcal{R}$	$\mathcal{P}$

2.1 Outcome Functions

If we wish to understand the outcome of a game, then we need only answer the basic question that any player would ask: *If I play first, can I force a win?* She need not worry at this point about her role playing second since, if that is the case, she will be taking on the role of "first" player after her opponent's move.

Definition 2.4. The *left-outcome function* maps positions to $\{\odot, \odot\}$: for a given position G,

$$o_L(G) = \begin{cases} \odot \text{ if Left can force a win moving first,} \\ \odot \text{ if Left cannot force a win moving first.} \end{cases}$$

Similarly, the *right-outcome function* is given by

$$o_R(G) = \begin{cases} \odot \text{ if Right cannot force a win moving first,} \\ \odot \text{ if Right can force a win moving first.} \end{cases}$$

Note that we color the Left and Right outcomes as a visual clue, though this is not a necessary part of the notation. The *outcome function* of a position G is $o(G) = (o_L(G), o_R(G))$.

The possible values for $o(G)$ are $(\odot, \odot)$, $(\odot, \odot)$, $(\odot, \odot)$, and $(\odot, \odot)$. Note that the first (or left) element of each ordered pair is a reference to Left's opinion when told that she is to move first, and the second (or right) element is a similar reference to Right's opinion if he is told that he is moving first. These correspond exactly with the outcomes $\mathcal{L}$, $\mathcal{N}$, $\mathcal{P}$, and $\mathcal{R}$ as can be verified by re-writing the tables that defined the outcome classes:

Class	Name	Outcome function
$\mathcal{N}$	Fuzzy	$o(G) = (\odot, \odot)$
$\mathcal{P}$	Zero	$o(G) = (\odot, \odot)$
$\mathcal{L}$	Positive	$o(G) = (\odot, \odot)$
$\mathcal{R}$	Negative	$o(G) = (\odot, \odot)$

and

Outcome classes	$o_R(G) = \odot$	$o_R(G) = \odot$
$o_L(G) = \odot$	$\mathcal{N}$	$\mathcal{L}$
$o_L(G) = \odot$	$\mathcal{R}$	$\mathcal{P}$

In the rest of this book, we will always take the result of the *outcome function*, $o(G)$, to be one of $\mathcal{N}, \mathcal{P}, \mathcal{L}, \mathcal{R}$ and not the ordered pair. As we are taking Left as positive and Right as negative (that is, $\mathcal{L} > \mathcal{R}$ for the left- and right-outcome functions), this naturally leads to the ordering for the outcomes given in Figure 2.1.

2.2   Game Positions and Options

Definition 2.5. A *game* (position) G is defined by its options, $G = \{\mathcal{G}^L \mid \mathcal{G}^R\}$, where $\mathcal{G}^L$ and $\mathcal{G}^R$ are the set of left and right options, respectively.

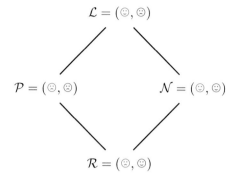

Figure 2.1. The partial order of the outcome classes and the corresponding values of the outcome function.

We usually omit the braces around sets $\mathcal{G}^L$ and $\mathcal{G}^R$ for brevity, so, in DOMI-NEERING for example,

$$\text{⊞} = \left\{\text{I}\!\boxminus, \boxminus\!\text{I} \mid \bowtie, \bowtie, \boxminus\right\}$$

We will often abbreviate a set of positions by simplifying and observing symmetries:

$$\text{⊞} = \left\{\text{⊟}, \text{⊟} \mid \text{□□}, \text{⊟}\right\}$$

In the last equation, we omitted several single cells in which no play is possible, and also □□□ since it has the same game tree as □□ .

The elements of $\mathcal{G}^L$ and $\mathcal{G}^R$ are called the left and right *options* of G, and we often write G^L or G^R to denote typical representatives of $\mathcal{G}^L$ and $\mathcal{G}^R$.

One, some, or all of the options of a game may be listed as games themselves:

$$\blacksquare\blacksquare\blacksquare\blacksquare\blacksquare = \left\{\blacksquare\blacksquare\blacksquare\blacksquare \mid \blacksquare\blacksquare\blacksquare\right\}$$
$$= \left\{\left\{\blacksquare\blacksquare\blacksquare \mid \blacksquare\blacksquare\right\} \mid \blacksquare\blacksquare\blacksquare\right\}$$

More iteration down the game tree could lead to many levels of braces. Another way to represent this game is to drop the internal braces and introduce a hierarchy of bars instead. Specifically, we might write

$$\blacksquare\blacksquare\blacksquare\blacksquare\blacksquare = \left\{\blacksquare\blacksquare\blacksquare \mid \blacksquare\blacksquare \;\middle\|\; \blacksquare\blacksquare\blacksquare\right\}$$

The outcome class of a game may be determined from those of its options instead of having to play out the whole game. This recursive approach is very useful when analyzing games, as we will see in the next section.

Observation 2.6. The outcome class of a game, G, can be determined from the outcome classes of its options as shown in the following table:

	Some $G^R \in \mathcal{R} \cup \mathcal{P}$	All $G^R \in \mathcal{L} \cup \mathcal{N}$
Some $G^L \in \mathcal{L} \cup \mathcal{P}$	$\mathcal{N}$	$\mathcal{L}$
All $G^L \in \mathcal{R} \cup \mathcal{N}$	$\mathcal{R}$	$\mathcal{P}$

Proof: We will first confirm the upper-right entry in the table.

If Left has an option in $\mathcal{L} \cup \mathcal{P}$ then Left has a winning move. If Right has only options in $\mathcal{L} \cup \mathcal{N}$ then Right has no winning move. Therefore, $G \in \mathcal{L}$.

Conversely, if $G \in \mathcal{L}$ then Left can win playing first. That is, she has a winning first move. This move is to a position in $\mathcal{G}^L$ from which she wins playing second. Thus, there is an option belonging to $\mathcal{G}^L$ that lies in $\mathcal{L} \cup \mathcal{P}$. Because Left can also win playing second, Right has no good first move and all Right's options must be in $\mathcal{L} \cup \mathcal{N}$.

The verifications of the other three entries in the table are similar and left as exercises for the reader.   □

Note that for a leaf of a game tree, i.e., a game with no options for either player, the statements "all $G^R \in \mathcal{L} \cup \mathcal{N}$" and "all $G^L \in \mathcal{R} \cup \mathcal{N}$" are both vacuously satisfied. All zero of the options are in the sets. So all leaves are in $\mathcal{P}$. Leaves of game trees are also sometimes called *terminal positions.*

We define the *positions* of G recursively to consist of G itself and the positions of the options of G.[1] In other words, the positions of G are those games that one can reach from G in zero or more moves, allowing for the possibility that one player may move more than once consecutively. For example, the positions of the DOMINEERING game are

Again, we will often abbreviate a set of positions by simplifying and observing symmetries. For instance, we could choose to write the positions of as

Example 2.7. The following game tree shows positions reachable from

but if two moves are symmetric, say

---

[1] Even though the positions of G include G itself, we will usually redundantly write "G and its positions" to be perfectly clear.

only one is included. Also, in this diagram, we removed any covered or singleton squares, writing these last positions as

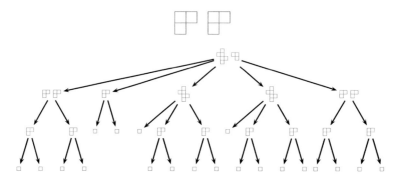

We can now identify each position's outcome class by referring only to the shape of the game tree using Observation 2.6. First, the leaves of the tree are $\mathcal{P}$-positions:

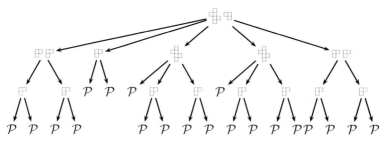

Now, continuing to work upward from the leaves using Observation 2.6, all the nodes of the tree can be classified according to their outcome classes:

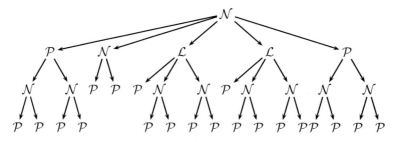

Exercise 2.8. Redo Example 2.7 using ▢▢▢.

Example 2.9. In a game of MAIZE (page 311), the outcome depends only on the location of the counter, so it's convenient to work out the outcome type by just filling in the grid starting from the terminal positions. For instance, to find the outcome of the following game of MAIZE

we can begin with the *leaves*:

Now looking for empty cells all of whose options are already labelled, we find

and continuing this process to the end gives

Exercise 2.10. Find the outcome of the previous example if the game is MAZE.

Example 2.11. The partizan subtraction game SUBTRACTION$(1, 2|2, 3)$ is played with a single heap of n counters. A move by Left is to remove one or two counters, while Right can remove two or three counters. Here, we wish to determine the outcome classes for every possible value of n.

Let G_n denote the game with n counters. Neither player can move from a heap of size 0, so $G_0 \in \mathcal{P}$. From G_1, Left has a move, while Right has none, so $G_1 \in \mathcal{L}$. Next, $G_2 \in \mathcal{N}$ for either player can remove two counters and win; $G_3 \in \mathcal{N}$ as well, for Left removes two counters and wins, while Right removes three counters and wins. Then, $G_4 \in \mathcal{P}$, for if the first player removes k, the second player can legally remove $4 - k$, since $1 + 3 = 2 + 2 = 4$.

One can quickly identify a pattern:

$$G_n \in \begin{cases} \mathcal{P} & \text{if } n \equiv 0 \pmod 4, \\ \mathcal{L} & \text{if } n \equiv 1 \pmod 4, \\ \mathcal{N} & \text{if } n \equiv 2 \text{ or } n \equiv 3 \pmod 4. \end{cases}$$

Proof:

- When $n \equiv 0 \pmod 4$, if the first player removes k, then the second can remove $4-k$, leaving a heap of size $n-4 \equiv 0 \pmod 4$, which by induction the second player wins. So, $G_n \in \mathcal{P}$ if $n \equiv 0 \pmod 4$.

- If $n \equiv 1 \pmod 4$, then Left playing first can remove one (winning by induction), but Right playing first only has moves to $n-2 \equiv 3$ and $n-3 \equiv 2 \pmod 4$, both of which leave a winning position for Left playing first. So, $G_n \in \mathcal{L}$ if $n \equiv 1 \pmod 4$.

- If $n \equiv 2 \pmod 4$, then either player can remove two leaving $n-2 \equiv 0 \pmod 4 \in \mathcal{P}$. So, $G_n \in \mathcal{N}$ if $n \equiv 2 \pmod 4$.

- Finally, if $n \equiv 3 \pmod 4$, then Left can remove two, leaving $n-2 \equiv 1 \pmod 4 \in \mathcal{L}$, or Right can remove three leaving $n-3 \equiv 0 \pmod 4 \in \mathcal{P}$. So, $G_n \in \mathcal{N}$ if $n \equiv 3 \pmod 4$.   □

Problem 6 identifies a more general reason why this example was periodic with period 4.

2.3   Impartial Games: Minding Your $\mathcal{P}$s and $\mathcal{N}$s

Definition 2.12. A game is *impartial* if both players have the same options from any position.

For example, GEOGRAPHY, NIM, and most subtraction games are impartial. CLOBBER, DOMINEERING, and CHESS are not impartial, since Left cannot move (or, in DOMINEERING, place) Right's pieces.

In the development of the theory, impartial games were studied first, and there are many impartial games that are currently unsolved despite many attempts over many years. NIM was analyzed by Bouton [Bou02] in 1902. In the 1930s and 1940s, Sprague and Grundy extended Bouton's analysis and showed that it applied to all impartial games. Guy continued the work throughout the 1950s, 1960s, and 1970s. So prevalent are impartial games that games which are not impartial have been dubbed *partizan* to distinguish them.

Impartial games form a nice subclass of games, and we will explore them more in Chapter 7. One property that makes this subclass interesting is that they only belong to two outcome classes.

Theorem 2.13. *If G is an impartial game then G is in either $\mathcal{N}$ or $\mathcal{P}$.*

Proof (by strategy stealing): If G were in $\mathcal{L}$ then Left could win going first, but then Right going first can use Left's strategy and win. □

Exercise 2.14. Give an inductive proof of Theorem 2.13 using Observation 2.6.

To classify the outcomes of an impartial game, the following observation is fundamental:

Theorem 2.15. (Partition Theorem for Impartial Games) *Suppose that the positions of a finite impartial game can be partitioned into mutually exclusive sets A and B with the following properties:*

- *every option of a position in A is in B; and*

- *every position in B has at least one option in A.*

Then A is the set of $\mathcal{P}$-positions and B is the set of $\mathcal{N}$-positions.

Proof: See Problem 3. □

Exercise 2.16. Explain why Theorem 2.15 correctly addresses the terminal positions of the game.

The mechanics of finding the outcome of an impartial position are easier than those of a partizan one. We again draw the game tree then label the vertices recursively from the leaves of the tree to the root, only this time the only labels are $\mathcal{P}$ and $\mathcal{N}$. (Whether or not this is easy, of course, depends on the depth and width of the tree.) For a normal play, impartial game, the terminal positions are all $\mathcal{P}$ and thereafter a position is labeled $\mathcal{P}$ or $\mathcal{N}$ according to the following rules:

- It is an $\mathcal{N}$-position if at least one of its options is a $\mathcal{P}$-position.

- It is a $\mathcal{P}$-position if all of its options are $\mathcal{N}$-positions.

Example 2.17. Consider the CRAM position

Considering symmetries, there are only two distinct first moves:

From either, the second player can reach the $\mathcal{P}$-position

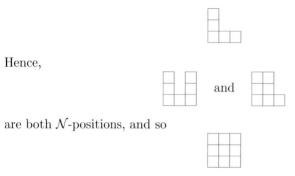

Hence,

and

are both $\mathcal{N}$-positions, and so

is a $\mathcal{P}$-position.

Exercise 2.18. (**Bottleneck Principle**) Let G be an impartial game with two options G_1 and G_2 (G could have other options) where the only option from G_2 is G_1. Show that G is an $\mathcal{N}$-position.

In partizan games, there isn't such an explicit result, but recognizing that a game has a bottleneck can give hints as to where the good moves are. For an example, see PARTIZAN ENDNIM in Section 2.6.

Of course, our goal is not to classify positions one by one, but rather to identify patterns that allow us to quickly determine whether any position of a game is in $\mathcal{P}$ or $\mathcal{N}$. Our opponent would have long since lost interest in playing us if we drew the game tree for every game we played.

Example 2.19. (CUTTHROAT STARS) A move in impartial CUTTHROAT consists of deleting a vertex and any incident edges from a graph with the proviso that at least one edge must be deleted on every move. We will play CUTTHROAT on star graphs. A *star graph* is a graph with one central node, any number of radial nodes, and one edge connecting the central node to each radial node:

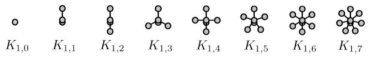

$$K_{1,0} \qquad K_{1,1} \qquad K_{1,2} \qquad K_{1,3} \qquad K_{1,4} \qquad K_{1,5} \qquad K_{1,6} \qquad K_{1,7}$$

In graph theory, a star graph with n radial nodes is denoted $K_{1,n}$. We will refer to n as the size of the star.

On a star of size 1 only a single move is available. On all larger stars there are two types of move: the *supernova,* which removes the central vertex, leaving a set of isolated vertices on which no further moves are possible; and the *shrink,* which removes an outer vertex, thus decreasing the size of the star by 1. From now on, all star sizes will be at least 1 — any isolated vertices left by a supernova move are not of any interest to us and are discarded.

A single star is always an $\mathcal{N}$-position, since the first player can simply use the supernova move. In multiple star positions, the $\mathcal{P}$-positions are precisely those where the number of even stars and the number of odd stars are both even.

To see this, we show how the partition of positions implicit in the above satisfies the conditions of Theorem 2.15:

- If there is an even number both of even stars and of odd stars, then a supernova move changes the parity of either the odd or the even stars, while a shrink move changes the parity of both. Note that a supernova move on an even star does not leave an even star since we ignore the resulting isolated vertices.

- If one type of star occurs with odd parity, then a supernova move on a star of that type makes the parities of both types even. If both types of star occur with odd parity, then a shrink move on an even star leaves an even number of both types.

If we wish to apply Theorem 2.15, then how do we find the right partition in the first place? A good starting point is to catalog a few small positions and identify a pattern. Then, use Theorem 2.15 (or an inductive argument, or both) to confirm the pattern. In Chapter 7 we present much more powerful tools for analyzing games that are disjunctive sums.

Let S be a set of positive integers. The game SUBTRACTION(S) is played with a heap of n counters, and a play is to select some $s \in S$ less than or equal to n and remove s counters, leaving a heap of size $n - s$.

Exercise 2.20. Denote by G_n the game SUBTRACTION$(1, 3, 4)$ played on a heap of n counters. Determine which G_n are $\mathcal{P}$-positions and prove your answer. (*Hint:* You should find that the game is periodic with period 7.)

2.4   Case Study: Roll the Lawn

The game of ROLL THE LAWN is played on a sequence of bumps (positive integers), non-bumps (zeros), and a roller. Left moves the roller leftward, and Right rightward, rolling out the bumps. Rolling over a bump decreases its size by 1 until it reaches 0.

For example, in the following position, Right moves first over two bumps, then Left moves back over one:

$$[5, \odot, 1, 2] \xrightarrow{R} [5, 0, 1, \odot] \xrightarrow{L} [5, 0, \odot, 0].$$

Left has won since Right has no move as there are no bumps to the right of the roller. Right could also have played over one bump, after which Left's move is

forced:
$$[5, \circ, 1, 2] \xrightarrow{R} [5, 0, \circ, 2] \xrightarrow{L} [\circ, 4, 0, 2].$$

Since all the remaining bumps are of even height, whatever move(s) Right makes now, Left can always move the roller back to the extreme left; therefore, Left will win.

Exercise 2.21.

(a) What is the outcome type of $[1, 1, 1, \circ, 1, 1]$?

(b) What about $[1, 1, \circ, 5]$?

(c) What if all the bumps are of height 2?

(d) And if the bumps are all of height 1?

The results of this exercise already suggest that parity is going to be an important part of the analysis of the outcome of a general ROLL THE LAWN position. Basically, bumps of even height are not really a resource for either player since a move across one of them in one direction always leaves a move back in the other. So, it seems natural to define the *bias*, $b(G)$, of a ROLL THE LAWN position G to be the difference between the number of bumps of odd size to the left of the roller and the number of bumps of odd size to the right of the roller. Note that we can freely ignore any bumps of height 0 in a ROLL THE LAWN position.

Theorem 2.22. *Let G be a* ROLL THE LAWN *position. Then*

$$o(G) = \begin{cases} \mathcal{R} & \text{if } b(G) < 0, \\ \mathcal{P} & \text{if } b(G) = 0, \\ \mathcal{L} & \text{if } b(G) > 0. \end{cases}$$

Proof: An equivalent form of the theorem is

$$o_L(G) = \begin{cases} \smiley & \text{if } b(G) > 0, \\ \frownie & \text{if } b(G) \le 0; \end{cases} \quad o_R(G) = \begin{cases} \smiley & \text{if } b(G) < 0, \\ \frownie & \text{if } b(G) \ge 0. \end{cases}$$

We note that a move by Left reduces the bias by 1 for every bump over which Left rolls. If it's an odd bump, it had contributed 1 to the bias, and after rolling it contributes 0. If it's an even bump, it had contributed 0 to the bias, and after rolling it contributes -1 because it ends up to the right of the roller. Similarly, Right's moves increase the bias by 1 per rolled bump. Furthermore, if a player has any legal move, the player can move so as to adjust the bias by 1 (by rolling over only one bump).

Because of the symmetry, it suffices to prove the first case, o_L. If $b(G) > 0$ before Left's move, Left will win by rolling over one bump, only leaving $b(G) \geq 0$ from which Right has no winning move (by induction). Similarly, if $b(G) \leq 0$, all of Left's moves leave $b(G) < 0$ and Right will win by induction. $\square$

The proof actually shows a little more than the claim — namely, if either player has a winning first move, then moving the roller one space is a winning move. Knowing this, the players can be very lazy and simply play out the game in that fashion until someone wins!

2.5   Case Study: Timber

In this case study we will analyse the outcome of a simple impartial game called TIMBER and see some surprising connections with a famous combinatorial sequence.

TIMBER is an impartial game played with dominoes in a straight line, each domino teetering, ready to fall over to the left or to the right. Each player can choose any domino but each domino can only be pushed (gently, perhaps) in one direction, which then topples all the other dominoes down the line (no matter in which direction they'd been teetering). For example, ▮▮▮ has the

options ▮▮ ,   , and ▮   , obtained by knocking over the first, last, and middle dominoes, respectively.

Throughout this section, we let Greek letters (α, β, ...) represent rows of dominoes in TIMBER positions. The position $\alpha\beta$ is obtained by placing the dominoes in arrangement α immediately followed by β, with no space in between them.

We first make a few simple observations:

- If $G = \alpha$▮ or $G = $▮$\beta$, then $o(G) = \mathcal{N}$ since the whole position can be knocked over with one move.

- If $G = \alpha$▮▮β, then the indicated ▮▮ form a *poison pair* — if either player knocks over one of them, the other player can win immediately by toppling the other.

- The smallest $\mathcal{P}$-positions are   , ▮▮, ▮▮▮▮, ▮▮▮▮, ▮▮▮▮▮▮,

The last of these observations leads to some interesting questions. Why do there not seem to be $\mathcal{P}$-positions with an odd number of dominoes? Why are the number of leftward-leaning and rightward-leaning dominoes equal in a $\mathcal{P}$-position? Is this a sufficient condition?

The last of these questions is easily dealt with: it is not the case that every position with the same number of dominoes leaning each way is a $\mathcal{P}$-position, since both ▨▨ and ▨▨▨▨▨▨ (and many other examples) are $\mathcal{N}$-positions.

The second observation is very important and leads to a characterization of the outcome classes and to a playable strategy. Consider a position $\alpha\,▨▨\,\beta$ — we claim that players who are trying to win can just ignore the poison pair completely and play as if the game were just $\alpha\beta$. That's because it's never a winning move to knock over a domino in the poisoned pair (though you may eventually be forced to), and any other move either destroys the poison pair (this happens if you knock over a ▨ in α or a ▨ in β), resulting in a position that's literally reachable from $\alpha\beta$ in one move, or leads to a position of the form $\alpha'\,▨▨\,\beta$ or $\alpha\,▨▨\,\beta'$ that inductively corresponds to a position reachable from $\alpha\beta$ by the corresponding move to $\alpha'\beta$ or $\alpha\beta'$.

Lemma 2.23. *The outcome type of* $\alpha\,▨▨\,\beta$ *is the same as that of* $\alpha\beta$.

Proof: This follows immediately from the previous paragraph by a simple inductive argument. If $o(\alpha\beta) = \mathcal{N}$ then the first player can win $\alpha\,▨▨\,\beta$ by making the move corresponding to a winning move in $\alpha\beta$ (by induction). On the other hand, if $o(\alpha\beta) = \mathcal{P}$ then the first player in $\alpha\,▨▨\,\beta$ has no good choices. If he moves on the poisoned pair, then the second player can win immediately. But, if he moves elsewhere, then the second player plays a winning response to the corresponding move on $\alpha\beta$ and (by induction) wins eventually.   □

Now, why stop with ignoring one poisoned pair? Using the preceding result the players can effectively ignore *all* the poisoned pairs in a string — and even more than that. Consider ▨▨▨▨. We know that the middle ▨▨ is poison and can be ignored, but that leaves the outermost ▨▨, another poison pair. In other words, the outer dominoes in the original position are "slow acting poison" — knocking one of them over is not immediately fatal but leads to death (or at least loss) eventually anyhow.

We call the elimination of a consecutive pair ▨▨ a *reduction*. The previous lemma and an observation based on the preceding discussion gives an immediate characterization of the $\mathcal{P}$-positions.

Theorem 2.24. *Game G is a $\mathcal{P}$-position if and only if G can be reduced to the empty position.*

Since a reduction eliminates one ▨ and one ▨, this characterization shows that all $\mathcal{P}$ positions are of even length and have the same number of ▨s as ▨s. How do we win as first player if the reductions *don't* lead to the empty string?

Well, what can a string that allows no reductions look like? It must not contain any ▎▎ pairs, so if it contains any ▎s, they must all be at the right-hand end of the string, and if it contains any ▎s, they must all be at the left-hand end of the string. That leads to the following:

Theorem 2.25. (Strategy for TIMBER**)** *Given a* TIMBER *position, successively pair off and eliminate poison pairs. If there are no dominoes remaining, then the original position is a* P*-position and we should invite the opponent to move first. If there are some dominoes left, the position will be m* ▎s *followed by n* ▎s, *where m or n might be 0 but not both. Toppling the domino of the original position corresponding to the leftmost of these* ▎s *or the rightmost of these* ▎s *will leave a* P*-position.*

Interestingly, any move other than the ones specified in the strategy is a losing one, i.e., there will always be exactly one (if the final reduced string consists entirely of ▎s or ▎s), two (if it contains both), or zero (if it contains neither) winning moves.

Exercise 2.26. What types of position are ▎▎▎▎▎▎ and ▎▎▎▎▎▎▎▎▎▎, and what are the winning moves (if any)?

Now we want to start again and come to an understanding of the outcome of TIMBER in another way. If you started computing P-positions by hand (or better, with the help of some software, e.g., CGSuite), you would soon observe that they all contained an equal number of left-leaning and right-leaning dominoes. Moreover, if you counted the number of P-positions of length $2n$ starting from $n = 0$, you would obtain a sequence beginning $1, 1, 2, 5, 14, 42, 132$. According to the On-Line Encyclopedia of Integer Sequences [Slo, Sequence A000108], another very famous sequence — the Catalan numbers — also begins this way. Could it be that there is a correspondence between the P-positions of TIMBER and one of the many types of combinatorial structures that are enumerated by the Catalan numbers? Of course it could.

Among other things, the Catalan numbers count the number of *Dyck paths*. A *Dyck path* is a sequence of segments in the Cartesian plane that start at $(0,0)$, end at $(2n, 0)$, at each step either make a $(+1, +1)$ step or a $(+1, -1)$ step, and never go below the x-axis. We will illustrate a connection with Dyck paths and use Theorem 2.15 to show that it is correct.

To each sequence of ▎s and ▎s we can associate a path beginning at $(0,0)$ where, for each ▎ we make a $(+1, +1)$ step, and for each ▎ a $(+1, -1)$ step. We claim that the P-positions of TIMBER are exactly those for which the corresponding path is a Dyck path. In fact, using the ideas of Section 1.6, we

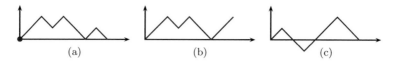

Figure 2.2. Dyck paths go from $(0,0)$ to $(2n,0)$ by making steps $(+1,+1)$ or $(+1,-1)$ without going below the x-axis. Figure (a) is a Dyck path, while (b) is not for it fails to end on the x-axis and (c) is not since it goes below the x-axis.

might as well consider a new (but equivalent) game that is played on paths of $(+1,+1)$ and $(+1,-1)$ steps (starting from $(0,0)$) where a move consists of deleting a $(+1,+1)$ step and all the preceding ones (and then shifting the resulting path so that it starts at $(0,0)$) or deleting a $(+1,-1)$ step and all the following ones. We will still think of these paths as being described by sequences of ⫽s and ⫽s (though Us and Ds would be more traditional).

According to Theorem 2.15, to prove the claim that in this game the $\mathcal{P}$-positions are exactly the Dyck paths, it suffices to show two things:

- no move from a Dyck path leads to another Dyck path, and

- for any position that is not a Dyck path, there is a move to a Dyck path.

For the first part, suppose that we have a Dyck path $\alpha \mathbin{⫽} \beta$ and consider deleting the prefix up to and including the designated ⫽. The remaining path β corresponds to the part of the original path beginning after the $(+1,+1)$ step represented by the ⫽. But, the y-coordinate at the end of that step in the original path was positive, so the new path finishes strictly below the level at which it begins and therefore is not a Dyck path. If we consider deleting a suffix beginning with a ⫽ from a Dyck path $\theta \mathbin{⫽} \tau$, then we note that the start point before the ⫽ was at a positive y-coordinate, and so θ is not a Dyck path.

For the second part, if a path begins with ⫽, we can delete the whole thing and leave the empty path (which is a Dyck path). If a path begins with ⫽ but is not a Dyck path, then one of two things must happen: at some point the y-coordinate is negative, or the final y-coordinate is strictly positive. In the first case, write the path as $\alpha \mathbin{⫽} \beta$, where the indicated ⫽ step is the one that first gives a negative y-coordinate. Then, α is a Dyck path and we can move to it by deleting the suffix beginning with the indicated ⫽. In the second case, consider the last step of the path that begins below the path's final height. This is necessarily a ⫽ step, the path is of the form $\theta \mathbin{⫽} \tau$, and τ is a Dyck path to which we can move by deleting the prefix $\theta \mathbin{⫽}$.

Exercise 2.27. Which of the two arguments characterizing the $\mathcal{P}$-positions of TIMBER do you prefer? Why?

2.6   Case Study: Partizan Endnim

Louise (Left) and Richard (Right) are fork-lift operators, with a penchant for combinatorial games. Many of the warehouses, from which they need to remove boxes, have the boxes in stacks with the stacks arranged in a row. Only boxes belonging to the stacks at the end of a row are accessible but the fork-lifts are sufficiently powerful that they can move an entire stack of boxes if necessary. The game that Louise and Richard play most often is won by the player who removes the last box from a row of stacks. In PARTIZAN ENDNIM analyzed in [AN01] and [DKW03], Louise must remove boxes from the leftmost stack and Richard is restricted to removing from the rightmost one.

For example, Left's legal options from 23341 are to 13341 or 3341, while Right has only one legal move, to 2334.

We use boldface Latin characters to stand for strings of positive integers, and non-bold characters for single positive integers. Concatenation of strings is denoted by their juxtaposition. So, for instance, we might write 23341 as $a\mathbf{w}b$ where $a = 2$, $\mathbf{w} = 334$, and $b = 1$.

Exercise 2.28. Build a table of outcome classes of $a\mathbf{w}$ for $a \in \{0, 1, 2, \ldots\}$ when

(a) $\mathbf{w} = 22$;

(b) $\mathbf{w} = 23$.

Exercise 2.29. Prove the following: if Left has a winning move from $a\mathbf{w}b$, then one of

- removing a single box or

- removing the entire stack

is a winning move.

Note that for a fixed, non-empty string $\mathbf{w}$, if a is large enough, Left can win from $a\mathbf{w}$. For example, if the left heap has more boxes than all the remaining heaps combined, then all Left need do is remove one box at a time until Right has removed all the boxes in $\mathbf{w}$, and then Left can remove the remaining stack. So, we can define

$L(\mathbf{w}) = $ the minimum $a \geq 0$ such that Left wins $a\mathbf{w}$ moving second, and

$R(\mathbf{w}) = $ the minimum $b \geq 0$ such that Right wins $\mathbf{w}b$ moving second.

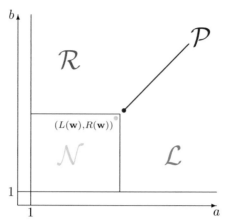

Figure 2.3. The outcome classes of the game awb for all $a, b > 0$. The green circle represents the triple point, $(L(\mathbf{w}), R(\mathbf{w}))$, which has an outcome class of $\mathcal{N}$. The filled circle is the point $(L(\mathbf{w}) + 1, R(\mathbf{w}) + 1)$, and the games on the line originating from the purple circle, which are of the form $(L(\mathbf{w}) + c, R(\mathbf{w}) + c)$, where $c > 0$, have an outcome class of $\mathcal{P}$.

Observation 2.30. We can use $L(\mathbf{w}b)$ and $R(a\mathbf{w})$ to determine the outcome class of awb. In particular, the outcome class of awb for $a, b > 0$ is given by

$$
awb \in \begin{cases}
\mathcal{L} & \text{if } a > L(\mathbf{w}b) \text{ and } b \leq R(a\mathbf{w}), \\
\mathcal{R} & \text{if } a \leq L(\mathbf{w}b) \text{ and } b > R(a\mathbf{w}), \\
\mathcal{N} & \text{if } a > L(\mathbf{w}b) \text{ and } b > R(a\mathbf{w}), \\
\mathcal{P} & \text{if } a \leq L(\mathbf{w}b) \text{ and } b \leq R(a\mathbf{w}).
\end{cases}
$$

Exercise 2.31. Prove the last observation.

In fact, we can do even better by determining the outcome class of awb using just $L(\mathbf{w})$ and $R(\mathbf{w})$. For fixed $\mathbf{w}$, there is one critical point of note, and that determines the outcome class of any awb!

Definition 2.32. The *triple point* of $\mathbf{w}$ is $(L(\mathbf{w}), R(\mathbf{w}))$.

Theorem 2.33. *The outcome class of awb for $a, b > 0$ is given by*

$$
awb \in \begin{cases}
\mathcal{N} & \text{if } a \leq L(\mathbf{w}) \text{ and } b \leq R(\mathbf{w}), \\
\mathcal{P} & \text{if } a = L(\mathbf{w}) + c \text{ and } b = R(\mathbf{w}) + c \text{ for some } c > 0, \\
\mathcal{L} & \text{if } a = L(\mathbf{w}) + c \text{ and } b < R(\mathbf{w}) + c \text{ for some } c > 0, \\
\mathcal{R} & \text{if } a < L(\mathbf{w}) + c \text{ and } b = R(\mathbf{w}) + c \text{ for some } c > 0.
\end{cases}
$$

This theorem is visually rendered in Figure 2.3.

Proof: First, assume that $a \leq L(\mathbf{w})$ and $b \leq R(\mathbf{w})$. If Left removes all of a, Right cannot win since any move is to $\mathbf{w}b'$ where $b' < b$ and b was the least value such that Right wins moving second on $\mathbf{w}b$. Symmetrically, Right can also win moving first by removing all of b. Thus, $a\mathbf{w}b \in \mathcal{N}$.

Next, assume that $a = L(\mathbf{w}) + c$ and $b = R(\mathbf{w}) + c$ for some $c > 0$. Also, assume that Left moves first. If Left changes the size of a to $a' = L(\mathbf{w}) + c'$ where $0 < c' < c$, Right simply responds by moving on b to $b' = R(\mathbf{w}) + c'$. By induction, this position is in $\mathcal{P}$. On the other hand, if Left changes the size of a to a' where $a' \leq L(\mathbf{w})$, Right can win by removing b as shown in the previous case. Thus, Left loses moving first. Symmetrically, Right also loses moving first. So, $a\mathbf{w}b \in \mathcal{P}$.

Finally, assume that $a = L(\mathbf{w}) + c$ and $b < R(\mathbf{w}) + c$ for some $c > 0$. If $b \leq R(\mathbf{w})$, Left can win moving first by removing all of a as shown in the first case. If $b > R(\mathbf{w})$, Left wins moving first by changing the size of a to $L(\mathbf{w}) + c'$, where c' is defined by $b = R(\mathbf{w}) + c'$, which is in $\mathcal{P}$ by induction. Left can win moving second from $a\mathbf{w}b$ by making the same responses as in the previous case. Thus, $a\mathbf{w}b \in \mathcal{L}$. □

So, if we can find the triple point, we can analyze the whole position. The functions $R(\cdot)$ and $L(\cdot)$ can be easily computed recursively:

Theorem 2.34.

$$R(a\mathbf{w}) = \begin{cases} 0 & \text{if } a \leq L(\mathbf{w}), \\ R(\mathbf{w}) - L(\mathbf{w}) + a & \text{if } a > L(\mathbf{w}); \end{cases}$$

$$L(\mathbf{w}b) = \begin{cases} 0 & \text{if } b \leq R(\mathbf{w}), \\ L(\mathbf{w}) - R(\mathbf{w}) + b & \text{if } b > R(\mathbf{w}). \end{cases}$$

Proof: As already shown in Theorem 2.33, Left loses moving first on $a\mathbf{w}$ where $a \leq L(\mathbf{w})$, so $R(a\mathbf{w}) = 0$. On the other hand, assume that $a > L(\mathbf{w})$. Again by Theorem 2.33, the least value of b that lets Right win moving second on $a\mathbf{w}b$ is $b = R(\mathbf{w}) + c = R(\mathbf{w}) + (a - L(\mathbf{w}))$. □

An algorithm to compute $R(\mathbf{w})$ and $L(\mathbf{w})$ using the above recurrence can be written to take $\Theta(n^2)$ time, where n is the number of heaps in $\mathbf{w}$.

Example 2.35. As an example, we will determine who wins from

$$3\ 5\ 2\ 3\ 3\ 1\ 9$$

when Left moves first and when Right moves first. Fix $\mathbf{w} = 52331$. We wish to compute $L(\mathbf{w})$ and $R(\mathbf{w})$ using Theorem 2.34. For single-heap positions,

$L(a) = R(a) = a$, because $R(a) = R() + (a - L()) = 0 + (a - 0) = a$. For two-heap positions, we have

$$R(ab) = \begin{cases} 0 & \text{if } a \leq b, \\ a & \text{if } a > b; \end{cases} \qquad L(ab) = \begin{cases} 0 & \text{if } a \geq b, \\ b & \text{if } a < b. \end{cases}$$

We can compute $L(52331)$ by first calculating $L(\mathbf{w})$ and $R(\mathbf{w})$ for each shorter substring of heaps:

$\mathbf{w}$	$L(\mathbf{w})$	$R(\mathbf{w})$
523	0	2
233	6	2
331	1	6
5233	1	0
2331	0	7
52331	2	12

For instance, $R(523) = R(23) + (5 - L(23)) = 0 + (5 - 3) = 2$.

For the original position $awb = 3\mathbf{w}9$, we have $3 > L(\mathbf{w}) = 2$ and $9 \leq R(\mathbf{w}) = 12$, and hence, by Theorem 2.33, 3523319 is an $\mathcal{L}$-position.

A small change to an individual heap size can have a large effect on the triple point; see Problem 10.

Problems

1. Find the outcomes of playing MAIZE and MAZE in each of the following two boards. You should determine the outcome class for every possible starting square for the piece.

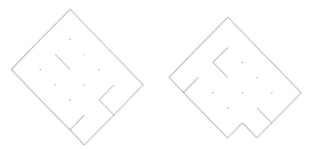

2. Find the sets A and B of Theorem 2.15 for

   (a) SUBTRACTION$(2, 3, 4)$;

   (b) SUBTRACTION$(1, 3, 6)$;

   (c) SUBTRACTION$(2, 3, 6)$;

(d) SUBTRACTION$(2^n : n = 0, 1, 2, \ldots)$;

(e) SUBTRACTION$(2^n : n = 1, 2, \ldots)$.

3. Give an inductive proof of Theorem 2.15 using Observation 2.6.

4. Consider the following two-player subtraction game. There is a heap of n counters. A move consists of removing any proper factor of n counters from the heap. (For example, if there are $n = 12$ counters, you can leave a heap with 11, 10, 9, 8, or 6 counters.) The player to leave a heap with one counter wins.

   (a) Determine a winning strategy from those positions in which you can win. In particular, you will need to determine which positions are $\mathcal{P}$-positions and which are $\mathcal{N}$-positions. Prove your answer by induction.

   (b) How about the misère version? In misère, the player to leave a heap with one counter loses.

5. GREEDY NIM is just like NIM, but you must always take from the (or a) largest heap. Identify the $\mathcal{P}$- and $\mathcal{N}$-positions in GREEDY NIM.

6. (This problem generalizes Example 2.11.) Fix a number p and a partizan subtraction game SUBTRACTION$(L \mid R)$, where $|L| = |R|$ and for each $x \in L$ there is a matching $y \in R$ such that $x + y = p$. (For instance, the game might be SUBTRACTION$(1, 2, 4, 7 \mid 2, 5, 7, 8)$ with $p = 9$.) Denote this game played on a heap of size n by G_n. Prove that the outcome classes of G_n are purely periodic with period p. That is, prove that for all $n \geq p$, G_n has the same outcome class as G_{n-p}.

7. (Modified CUTTHROAT STARS) Consider a collection of stars $K_{1,n}$. A move consists of deleting a vertex and the incident edges. Any vertex may be deleted so long as at least one star, somewhere, contains an edge before the move. In other words, players can play on isolated vertices unless all edges are gone. An isolated vertex will be considered a *trivial star,* other stars will be called *real stars.*

   Show that the $\mathcal{P}$-positions are those in which there are at least two real stars and the number of even stars is even.

8. The COMMON DIVISOR game is played with heaps of counters. A move is to choose a heap and take away a common divisor of all the heap sizes. For this game, $\gcd(0, a) = a$. An example game would be

$$(2, 6) \to (2, 4) \to (2, 3) \to (1, 3) \to (0, 3) \to (0, 0).$$

   Find the $\mathcal{P}$-positions in the two-heap game. (*Hint:* binary.)

9. Determine some of the $\mathcal{P}$-positions in two-dimensional CHOMP. In particular, determine *all* the $\mathcal{P}$-positions for width 1 and width 2 boards, and find at least two $\mathcal{P}$-positions for boards that include the following six squares:

10. (a) For $\mathbf{w} = 35551$, compute $L(\mathbf{w})$ and $R(\mathbf{w})$.

    (b) Repeat with $\mathbf{w} = 35451$.

11. Prove that in PARTIZAN ENDNIM there are no $\mathcal{N}$-positions of even length (meaning an even number of non-empty stacks).

12. Find all the winning moves for TIMBER, if any, in

    (a) ,

    (b) , and

    (c) .

13. The Catalan numbers also correspond to the number strings of $2n$ open- and closed-parentheses that are well-balanced. For example, ()(()) and (())((())) are well-balanced while)((()) and (())) are not. How would you connect this view of Catalans to the game of TIMBER?

14. Find a characterization of the misère $\mathcal{P}$-positions of TIMBER.

15. (Harder) Suppose that we add green dominoes (which can topple in either direction) to TIMBER. Characterize the $\mathcal{P}$-positions of this new game.

Preparation for Chapter 3

To the instructor: This chapter and the prep problems before the next chapter, motivate abstract algebraic thinking. Feel free to remind students of other groups or partial orders that they might have seen in your curriculum.

Chapter 3

Motivational Interlude: Sums of Games

It is a mistake to try to look too far ahead.

Sir Winston Churchill

3.1 Sums

So far, we have seen a collection of ad hoc techniques for analyzing games. Historically, that was the state of the art until Sprague [Spr35] and Grundy [Gru39] proved that Bouton's analysis [Bou02] of NIM could be applied to all impartial games.

Berlekamp, Conway, and Guy then set out to develop a unified theory of partizan games. They observed that many games have positions that are made up of independent components. Sometimes, as in NIM, the components are simply part of the game. In NIM, a player may play on any one of a number of independent heaps. Other games naturally split into components as play proceeds. Most famously, the endgame of the ancient and popular Asian game of GO has this property. We have seen this phenomenon in other games as well; in DOMINEERING, for instance, the playing field is often separated into different regions of play as in the following game:

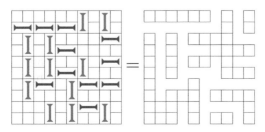

59

Given the fact that many games separate into independent regions, the natural question a mathematician asks is, "How can we exploit the decomposition?" In other words, what is the minimum amount of information that we need to know about an individual component to know how it behaves in the context of other components?

This naturally leads to a definition of *game sum:*

Definition 3.1. A *sum* of two or more game positions is the position obtained by placing the game positions side by side. When it is your move, you can make a single move in a summand of your choice. As usual, the last person to move wins.

For example, in DOMINEERING, we have

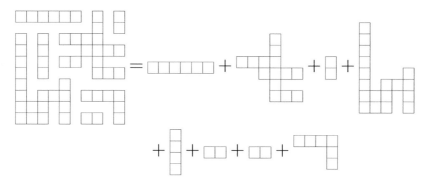

John Conway had a brilliant insight. Since game sums appear to be central to so many games, he chose to distill a few game axioms that encapsulate the following natural notions:

- A move is from a position to a subposition; the last player to play wins.

- In a sum of games, a player can move on any summand.

- In the negative of a game, the players' roles are reversed.

- Position A is at least as good as B (for Left) if Left is always satisfied when B is replaced by A in any sum.

We will list the formal axioms in the next chapter. The goal here is to motivate these axioms and explore some of their consequences.

This approach using "sums" has been so successful that it deserves a name and is sometimes referred to as *Conway Game Theory* or, more recently, *Additive Game Theory*. The theory develops techniques that are useful across a large general class of games. These techniques may or may not help with combinatorial games that do not break up into sums. HEX, for example, is a *maker-maker* game — both players are trying to create a path. The techniques used to analyze such games tend to be specific to the game rather than general; see Beck [Bec06] for example.

First, note what is *not* part of Conway's axioms. For example, games like GO and DOTS & BOXES appear to be excluded. Although they break up into sums, it is the person who accumulates the highest score who wins, not the last player to play. Conway chose to use a minimal number of axioms in order to develop a powerful and general mathematical foundation. What is remarkable is that, as we will see, a notion of score naturally develops from these axioms, making the theory applicable to games with scores.

Conway's axioms also appear to exclude misère games in which the last player to play *loses*. There are many misère games for which this proves to be only a minor inconvenience, but most misère games seem to be much harder than their non-misère counterparts. In just the last couple of years, tremendous progress has been made in understanding misère games; we briefly discuss this progress in Chapter ω.

The description of the axioms is suggestive of both an algebraic structure and a partial order. It includes references to addition, negation, and comparison. For the rest of the book, we will have two goals: to understand the structure implied by Conway's axioms, and to learn how to apply what we know about the structure to a wide variety of games.

3.2 Comparisons

In a sum of DOMINEERING games, Right will be satisfied if a □□ summand is replaced by □□□ since he gets one move in either of the summands and Left has the same options as before. Right will not only be satisfied but will be very happy if a □□ summand is replaced by □□□□ since now he will have the opportunity to make an extra move. The position □□□ is as good as □□, but □□□□ is better than either of these from Right's perspective.

Like Right, Left will be satisfied if a

summand is replaced by

or

It makes sense to say that both the

and

games are worth 1 move to Left; that

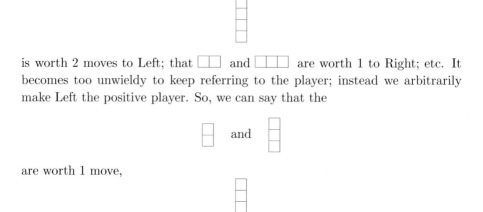

is worth 2 moves to Left; that ▢▢ and ▢▢▢ are worth 1 to Right; etc. It becomes too unwieldy to keep referring to the player; instead we arbitrarily make Left the positive player. So, we can say that the

and

are worth 1 move,

is worth 2, while their *negatives,* ▢▢ , ▢▢▢ , and ▢▢▢▢ , are worth −1, −1, and −2 moves, respectively.

In the rest of the book, we will be trying to find good "names" for the values of games, but the integers are easy and so we present them here. The zero game is $0 = \{\cdot|\cdot\}$ in which neither player has a move. To represent the positive integer n, we want Left to have n moves and Right to have none. So, Right's options are clear. After Left moves she should have $n - 1$ moves remaining, thus we define $n = \{n - 1|\cdot\}$. Correspondingly, $-n = \{\cdot| - (n - 1)\}$. Formally, we have the following definition:

Definition 3.2. The *integers* (as games) are defined recursively as follows:

- $0 = \{\cdot|\cdot\}$; and

- for a positive integer n, $n = \{n - 1|\cdot\}$ and $-n = \{\cdot| - (n - 1)\}$.

Having names for the integers as games also allows us to consider more complicated games such as $\{2| - 2\}$ or $\{3 \mid \{2|1\}\}$ — in the latter each player has a single option and the left option is a position worth 3 moves to Left while the right option is the position $\{2|1\}$.

In arithmetic, adding 0 to anything does not change the value. In DOMI-NEERING do the two summands 2×1 and 1×3 really cancel out just as $1 - 1 = 0$? To see whether or not this is true, consider how the players might respond if both are added to a game G. The goal is to show that whoever could have won G will also win in the new sum.

If Left could win G then she can play one of her winning strategies in G, ignoring the two new summands unless Right makes a move in one of them. At some point in the play, even as late as after the play in G has been exhausted,

Right will make a move outside of G. His only possible move of this type is to play in 1×3. Left can answer this move in 2×1, and Right must continue back in G as if this exchange had never happened. So, the outcome of the sum will still be a win for Left.

Similarly, if Right can win in G then likewise he can win in G plus these two new summands. Adding this particular 0 to a game does not change the outcome of the game! This argument, extended, also works for any second-player-win game H. Though we're being informal in this chapter, it's worth recording this fact formally.

In the sum $G + H$, a *One-Hand-Tied Principle* strategy is for the second player to *respond locally*. That is, always respond in the component, G or H, in which the opponent has just played. This can be a useful (but not always the very best) strategy. Before attempting the proof, it is worthwhile re-reading Section 2.1.

Lemma 3.3. *Let G and H be games. If $o_L(G) = o_L(H) = \odot$, then $o_L(G+H) = \odot$. If $o_R(G) = o_R(H) = \odot$, then $o_R(G + H) = \odot$.*

Proof: If Right can win both G and H playing second, then Right has only to respond locally in $G + H$ to win going second. The same argument works if Left can win both going second. □

Proposition 3.4. *Let G be any game and let $Z \in \mathcal{P}$ be any game (that is, Z is a second-player win). Then $o(G) = o(G + Z)$.*

Proof: If $o_L(G) = \odot$ then Left, playing first, has winning moves in G to some G^L. Now $o_R(G^L) = o_R(Z) = \odot$, so by Lemma 3.3, Left can respond locally to force a win; that is, $o_L(G + Z) = o_L(G) = \odot$. If $o_L(G) = \odot$ then, in $G + Z$, Right responds locally and forces a win; that is, $o_L(G+Z) = o_L(G) = \odot$. The same arguments with Left and Right switched give $o_R(G + Z) = o_R(G)$. The string of equalities

$$o(G + Z) = (o_L(G + Z), o_R(G + Z)) = (o_L(G), o_R(G)) = o(G)$$

gives the result. □

Furthermore, if Z is any game that has the property that for any game G we have $o(G) = o(G + Z)$, then in particular $o(\{\cdot \mid \cdot\}) = o(\{\cdot \mid \cdot\} + Z)$ and the sum of the empty game and Z (which is just Z, obviously) are the same, and so Z must be a second-player win. Thus, it makes sense to say:

Any second-player-win game has value 0, and no other games have this value.

A more careful reading of the proof of Proposition 3.4 shows that if Left wants to win in $G + Z$, she need only exploit the fact that she wins G and wins moving second on Z. Consequently, we know the following:

Proposition 3.5. *Let G be any game and let $X \in \mathcal{P} \cup \mathcal{L}$ be any game that Left wins moving second. The outcome class of $G + X$ is at least as favorable for Left as that of G; that is, $o(G + X) \geq o(G)$.*

Of course, there is also a dual version of this proposition, which applies to Right. Using Proposition 3.5 and its dual, we can determine an addition table that shows the possible outcome class of a sum given the outcome classes of the two summands:

	$\mathcal{L}$	$\mathcal{P}$	$\mathcal{R}$	$\mathcal{N}$
$\mathcal{L}$	$\mathcal{L}$	$\mathcal{L}$	?	$\mathcal{L}$ or $\mathcal{N}$
$\mathcal{P}$	$\mathcal{L}$	$\mathcal{P}$	$\mathcal{R}$	$\mathcal{N}$
$\mathcal{R}$	?	$\mathcal{R}$	$\mathcal{R}$	$\mathcal{R}$ or $\mathcal{N}$
$\mathcal{N}$	$\mathcal{L}$ or $\mathcal{N}$	$\mathcal{N}$	$\mathcal{R}$ or $\mathcal{N}$	?

In DOMINEERING, that 4×1 is actually better for Left than 2×1 is clear, but how would we test it? One way is to give the advantage of the 2×1 game to Right and to play the two games together. That is, play 4×1 plus 1×2. Left going first moves to 2×1 plus 1×2, which is a second-player win with Left as the second player. If Right moves first in the original game, he moves to 4×1, again a Left win. So, $(4 \times 1) + (1 \times 2) \in \mathcal{L}$. Not only intuitively, but also by the play of the game, the advantage to Left in 4×1 is greater than that in 2×1.

Can positions be worth a non-integer number of moves? There are certainly some candidates. Let

$$G = \text{[L-tromino]} \quad \text{and} \quad H = \text{[J-tromino]}$$

and consider the games

$$G + G + \text{[vertical domino]} \quad \text{and} \quad H + H + \text{[vertical domino]}$$

It seems clear that Left will not want to take the move available in [vertical domino] unless she has to. In

$$\text{[L-tromino]} + \text{[L-tromino]} + \text{[vertical domino]}$$

Left going first moves to

$$\text{[L-tromino with Left piece]} + \text{[L-tromino]} + \text{[vertical domino]}$$

and Right responds with

$$\text{[L-tromino with Left piece]} + \text{[L-tromino with Right piece]} + \text{[vertical domino]}$$

Now, both Left and Right have exactly one move each remaining with Left to play, so she loses. Right going first will move to

and Left will respond with

Both players have one move left but it is Right's turn to play and now he will lose. So, apparently, $G + G + 1 = 0$.

Similarly, in

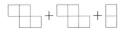

Left going first moves to

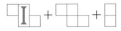

and Right responds with

Both Left and Right have exactly one move each remaining with Left to play so she loses. Right going first will move to

and Left will respond with

Both players have one move left but it is Right's turn to play and now he will lose. So, likewise, $H + H + 1 = 0$.

It seems that a good guess would be that both G and H are worth $-\frac{1}{2}$ moves. However, the arithmetic of games turns out to be more complicated than that. For one thing, G and H belong to different outcome classes, for G is a Right win and H is a first-player win. By our convention only G should be negative. Right would be happy to have G replace a second-player-win game (i.e., a value 0 game). So $G < 0$. Right *could* be unhappy if 2×1 were replaced by G. To show this, let's hand the 2×1 over to Left, which now becomes a 1×2, and compare. In

Left, going first, moves to

and wins. Right, going first, moves to

which Left wins, or he moves to

which is even worse since Left has two free moves. Therefore, $-1 < G < 0$, which together with $G + G = -1$ suggests that G is a good candidate for a half-move advantage to Right.

We will build techniques for doing this algebra of games more efficiently in the next chapter, and we will then be able to better formalize what it means for a position to be worth, say, a non-integer number of moves in Chapter 5, which introduces all sorts of game *values,* not merely numbers.

Exercise 3.6. Determine the outcome class of

$$G + H + \boxed{} = \boxed{} + \boxed{} + \boxed{}$$

3.3   Equality and Identity

When are two games equal? It seems a silly question — and perhaps it is if you interpret "equal" as "identical." Two CHESS games in which the same first four moves have been played are certainly equal (at this point) under this definition. But what if we reach identical CHESS positions by different sequences of moves (called a *transposition*)? Certainly, from the standpoint of the game as it will proceed from here, the two games must still be considered equal. On the human side, though, they may not be equal at all — there may be important psychological and tactical messages that can be read from the actual sequence of moves chosen. However, as theorists of games, we will not even pretend to address such messages, so for us these two games would still be identical (and hence certainly equal).

Can two different games ever be equal? Consider the DOMINEERING position

and the NIM position consisting of a single counter. In both these positions each player has precisely one legal move and after making that move his or her opponent will have no move available. So, these two games have identical (and rather trivial) game trees — there is an exact correspondence between any sequence of moves from one game and any sequence of moves from the other. So, from anything except an aesthetic standpoint, the two games are the same and can be considered equal. Moreover, the game tree captures completely all the possible sequences of play in the game — the ebbs and flows of fortune, the tricks that might be tried, the seemingly advantageous moves that are really terrible traps. Thus, the idealist might well choose to argue that two games should be considered equal if and only if they have identical game trees.

The utter pragmatist looks at games differently. To him, when faced with a game, the only question of interest is, "Who will win?" That is, the sole significant feature of a game is its outcome class. Beyond the four possible outcome classes, everything is just window dressing, a way to hide the essential nature of the underlying position. Such a person might well claim that two games are equal if (and only if) they have the same outcome class.

Combinatorial game theory steers a middle course between these two positions. Though the ultimate goal in practice may well be to determine the outcome class of a particular game, as we do for some instances of DOMINEER-ING rectangles in the next section, the aim of the subject is to produce practical and theoretical tools, which enhance our ability to make such determinations. The problem with the purely pragmatic view is that it treats each game in isolation — it becomes a matter of brute-force search, possibly guided by some good heuristic techniques, to actually work out the relevant outcome class. What we would like to be able to do is to use theory to simplify this process.

A key insight in this simplification is the recognition that two games might as well be considered equal if they can be freely substituted for one another in any context (which we will take, without terribly strong justification, to mean *sum*) without changing the outcome class. This provides us with the outline of a strategy for analyzing complex games — decompose them into sums, and replace complicated summands by simpler ones to which they are equal. For instance, we have already seen that *any* second-player win can be ignored in a sum — it has no effect on the outcome class. Likewise, we have seen that the domineering position

behaves *just like*

If we had a reasonably large catalog of simplest forms of various DOMINEERING positions, then we would be well on our way to understanding how to play even quite complex DOMINEERING games perfectly.

In the next chapter we will pursue the goal of defining appropriate notions of sum, equality, and comparison for games in a way that takes advantage of the observations that we have made in this chapter. However, to demonstrate that even the pragmatist can draw some benefit from an understanding of the importance of decomposition in games, the next section shows how a player could use imagined or virtual decompositions as a guide to playing well in some (admittedly rather simple) games of DOMINEERING.

3.4   Case Study: Domineering Rectangles

A complete analysis of the game of DOMINEERING is not yet feasible, though there are techniques (many covered in this book) to help analyze a great many positions.

However, if we only wish to know the outcome classes, then DOMINEERING can be solved completely on various rectangular boards. Some classes are analyzed in [LMR96](also see [Ber88]). We present our version of their analysis of some $2 \times n$ positions, and all the $3 \times n$ positions. If we think of n as being large (after all, most positive integers are!), then it is intuitively clear that the $3 \times n$ positions should be advantageous to Right who plays the horizontal dominoes because the board can contain $3\lfloor \frac{n}{2} \rfloor$ horizontal dominoes but only n vertical ones.

Our analysis of these positions uses a powerful technique that simplifies many arguments. Here it is:

> If it is true that Right wins some game G when he promises not to make certain types of moves, then he has a winning strategy in G itself.

The point of this technique is that by enforcing a promise not to make certain kinds of moves, it may well be the case that the game becomes much simpler — but if this does not harm Right's winning chances, then it is at no cost to him. We call this the *One-Hand-Tied Principle* since Right is promising that he can win even with one hand tied behind his back.

In considering DOMINEERING played on rectangular boards, the types of promises that we will make on Right's behalf are not to play any move that crosses certain vertical dividing lines. That is, Right will promise to play as if the board were split along certain vertical lines into a sum of games. If this promise does not harm his winning strategy, then we can be sure that the original game is also a win for Right.

We first consider some $2 \times n$ cases:

$$\boxed{} \in \mathcal{R}.$$

To see this we first illustrate winning replies to each of Left's two possible first moves (up to symmetry):

We must also demonstrate a winning move for Right as first player:

This wins because, up to symmetry, Left has only one move available and then Right can end the game in the position

In the remaining arguments we will mostly leave the last step or two of the verification to you!

Now the One-Hand-Tied Principle implies that, for all $k \geq 1$, DOMINEERING played on a $2 \times 4k$ rectangle is in outcome class $\mathcal{R}$. Right will simply promise to pretend that the game is being played on k copies of 2×4 and use his winning strategy in each of them. A simple analysis shows that the 2×3 game is in $\mathcal{N}$. Then, applying the One-Hand-Tied Principle again, it follows that Right can always win the $2 \times (4k+3)$ games playing first (by making his winning move in the 2×3 game to begin with), so the $2 \times (4k+3)$ game is either in $\mathcal{N}$ or in $\mathcal{R}$. In fact, though we will not show it here, all the games $2 \times \{7, 11, 15, 19, 23, 27\}$ are in $\mathcal{N}$ while $2 \times (31 + 4t)$ is in $\mathcal{R}$ for any $t \geq 0$.

Suppose that a trustworthy mathematician told us that the 2×13 game was in $\mathcal{P}$ (this is, in fact, true). Could we make use of this information? Tying one of his hands, Right can afford to ignore any 2×13 parts of the game (planning to play the winning strategy as second player there if Left makes a move). Now observe that

$$36 = 9 \times 4;$$
$$37 = 6 \times 4 + 13;$$
$$38 = 3 \times 4 + 2 \times 13;$$
$$39 = 3 \times 13.$$

It follows that $2 \times \{36, 37, 38\} \in \mathcal{R}$ and that 2×39 is either in $\mathcal{R}$ or $\mathcal{P}$. But, as $39 = 9 \times 4 + 3$, we already knew that 2×39 is in $\mathcal{R}$ or $\mathcal{N}$. Therefore, it is in $\mathcal{R}$. Now, however, we have four consecutive values of k such that the $2 \times k$ game is in $\mathcal{R}$. By adding suitable copies of 2×4, this is also the case for all larger values of k. In fact, it turns out that the 2×27 game, which is in $\mathcal{N}$, is the last $2 \times n$ instance of DOMINEERING that does not belong to $\mathcal{R}$, but the detailed arguments required to show this are a bit too complex for us here.

Somewhat surprisingly it turns out that it is easier to analyze the $3 \times n$ positions. Clearly $3 \times 1 \in \mathcal{L}$, and we observed above that $2 \times 3 \in \mathcal{N}$, from which it follows that $3 \times 2 \in \mathcal{N}$ also (by symmetry). The 3×3 game is symmetrical and so must be in $\mathcal{N}$ or $\mathcal{P}$. However, it is clear that

$$\boxed{\;\vphantom{I}\text{I}\;} \in \mathcal{L},$$

and so the 3×3 game is in $\mathcal{N}$. We will show that $3 \times \{4, 5, 6, 7\} \in \mathcal{R}$, and then it follows as above that $3 \times k \in \mathcal{R}$ for all $k \geq 4$ using the One-Hand-Tied

Principle. In analyzing these four specific cases, we will frequently make use of symmetry to limit the number of moves that we need to consider, without explicitly mentioning it. Also, Right generally plays naively anywhere he can in the middle row unless we specify otherwise.

Consider first the 3×4 case. Suppose that the first pair of moves (regardless of who moves first) leads to

Right can tie one of his hands by vertically separating the board in two places leaving

$$\square + \square + \square\square + \square\square$$

which is in $\mathcal{R}$; Right will win whether he moved first or second. If, on the other hand, Left's first move is on an edge of the board, it is also easy to construct a winning strategy for Right.

For 3×5, Left's first move on an edge is bad by the previous argument. If Left's first move (as first or second player) is not in the middle column, Right can then tie his hands and separate the board into

$$\boxed{\text{I} \longmapsto} \;\Rightarrow\; \square \;+\; \square\square \;+\; \square\square \;+\; \square$$

which is also a Right win, for we argued in Section 3.2 that

$$\square \;+\; \square\square \in \mathcal{R}$$

and, of course,

$$\square \;+\; \square\square \in \mathcal{P}.$$

If Left moves first in the middle column, Right can respond as shown below:

Now Left has at most two moves remaining while Right has at least three (the two on the right hand side of the board and at least one in the 3×2 rectangle on the left), and so Right will win.

For 3×6 and 3×7, an initial move by Right in the middle row of the rightmost two columns leaves him with two more moves there, together with a position that he wins as second (or first) player, so that can't be bad!

So, for these boards we only need to demonstrate (and check) winning replies to Left's possible first moves (and moves by Left in the first column are known to be bad). For 3×6, Right immediately ties one hand to

and proceeds to play to maintain 90° rotational symmetry between the two boards.

For 3×7, Right's response to Left's first move is in the middle row at one end of the board. Even with the first two moves on the remaining 3×5, Left cannot prevent Right from getting two more moves there. For example, from the 3×5

Right plays the lower left and will get another move, and from

Right will get a move on both the top and bottom. In total, Right is guaranteed five moves throughout the game, enough to lock in a victory, because Left has at most five columns available after Right's initial move.

Problems

1. Who wins in each of these two sums of MAIZE positions?

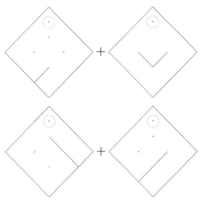

2. To which outcome class does the following sum of TOPPLING DOMINOES positions belong?

3. Prove that Left can win from the following DOMINEERING position moving first or second:

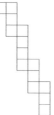

4. Let j and k be non-negative integers. What is the outcome of j copies of ☐☐ and k copies of ⊞ ?

Preparation for Chapter 4

Prep Problem 4.1. Show that if Left wins moving second on G_1 and G_2, Left can win moving second on $G_1 + G_2$.

Prep Problem 4.2. Give an example of games G_1 and G_2 such that Left can win moving first on G_1 and on G_2, but cannot win on $G_1 + G_2$, even if she can choose whether to move first or second on the sum.

Prep Problem 4.3. List the properties (or axioms) you usually associate with the symbols $=$, $+$, $-$, and $\geq$. One property might be, "If $a \geq b$ and $b \geq c$, then $a \geq c$"; that is, transitivity of $\geq$. After spending about 10 to 15 minutes listing as many as you can think of, compare your notes with a classmate.

To the instructor: In an undergraduate class, consider covering material from this chapter interleaved with examples from the next chapter to motivate this more theoretical chapter. In particular, this chapter provides axioms for combinatorial game theory and important theorems that can be derived reasonably directly from the axioms. The next chapter, however, places the axioms in context. (We chose this approach in order to maintain axiomatic continuity.) For a more example-driven development, present the games of HACK-ENBUSH and CUTCAKE as in Chapter 2 of *WW* [BCG01] through their section entitled "Comparing Hackenbush Positions," and assign problems primarily from the next chapter.

Chapter 4

The Algebra of Games

The grand aim of all science is to cover the
greatest number of empirical facts by
logical deduction from the smallest number
of hypotheses or axioms.

Albert Einstein

4.1 The Fundamental Definitions

We will state John Conway's definitions of the principal concepts of combinatorial game theory forthwith and then discuss and explain each of them in
sequence.[1] You should not expect to understand what these definitions mean
until you read the following subsections that discuss each one in greater detail.
Here are the five concepts to be defined: what a game is, what the sum of two
games is, what the negative of a game is, when two games are equal, and when
one game is preferred over another by Left.

$$G \quad \text{is} \quad \{\mathcal{G}^L \mid \mathcal{G}^R\}, \text{ where } \mathcal{G}^L \text{ and } \mathcal{G}^R \text{ are sets of games.} \quad (4.1)$$

$$G + H \quad \overset{\text{def}}{=} \quad \{\mathcal{G}^L + H,\ G + \mathcal{H}^L \mid \mathcal{G}^R + H,\ G + \mathcal{H}^R\}. \quad (4.2)$$

$$-G \quad \overset{\text{def}}{=} \quad \{-\mathcal{G}^R \mid -\mathcal{G}^L\}. \quad (4.3)$$

$$G = H \quad \text{if} \quad (\forall X)\ G + X \text{ has the same outcome as } H + X. \quad (4.4)$$

$$G \geq H \quad \text{if} \quad (\forall X)\ \text{Left wins } G + X \text{ whenever Left wins } H + X. \quad (4.5)$$

In (4.4) and (4.5) the "X" in question ranges over all games. Also, in (4.5)
we write "whenever" to mean $o_L(G + X) \geq o_L(H + X)$ and $o_R(G + X) \geq o_R(H + X)$ — in words, if Left wins moving *first* on $H + X$ then Left wins

---

[1]Conway actually defines $=$ and $\geq$ a bit differently, but these definitions are equivalent.

moving *first* on $G + X$ **and** if Left wins moving *second* on $H + X$ then Left wins moving *second* on $G + X$.

Recall the discussion of Section 3.3 and remember that at this point we need to be particularly careful to not think "$G = H$" (in the sense defined above) means that "G is the same as H." In fact, beginning on page 80, we will prove a number of theorems that tell us just how safe it is to make this identification. We say G is *isomorphic* to H, $G \cong H$, if G and H have identical game trees. As discussed in Section 3.3, if $G \cong H$, then to all intents and purposes G and H are the same game, and we can replace G by H (or vice versa) wherever it might occur.

Definition of a game

Effectively, we defined a game to be a pair of sets of games when we said that G was $\{\mathcal{G}^L \mid \mathcal{G}^R\}$. The intention of this definition is just what we saw previously in Section 2.2 — $\mathcal{G}^L$ represents the set of left options of G and $\mathcal{G}^R$ its right options. The definition itself is rather terse and recursive. The recursive part might — rather, *should* — trouble you, since we neglected to give a base case for the recursion. Thus, defining a game as a pair of sets of games sounds circular. But don't panic. We can *bootstrap* the definition since the sets could be empty. In particular, the definition tells us that $G \overset{\text{def}}{=} \{\mathcal{G}^L \mid \mathcal{G}^R\}$ where $\mathcal{G}^L = \mathcal{G}^R = \emptyset$ is a game, a game which we now dub *zero*. One advantage of a recursive definition without explicit base cases is that many inductive proofs based on the definition will not need explicit base cases either. (For a proof without a base case — but with an explanation for why it is not needed — see Theorem A.3 on page 261.)

Definition 4.1. The *birthday* of a game $G = \{\mathcal{G}^L \mid \mathcal{G}^R\}$ is defined recursively as 1 plus the maximum birthday of any game in $\mathcal{G}^L \cup \mathcal{G}^R$. For the base case, if $\mathcal{G}^L = \mathcal{G}^R = \emptyset$, then the birthday of G is 0.

In other words, the birthday of a game is the height of its game tree. A game is *born by day* n if its birthday is less than or equal to n. Each game has several birthdays, for example, the game $0 = \{\mid\}$ has birthday 0 and birthday 1 since 0 also equals $\{-1 \mid 1\}$. One goal of this chapter is to show that among all the games that are equal to some game G, there is a unique game with the least birthday. Since that game can replace G in any disjunctive sum, induction can then be based on birthdays.

The game $0 \overset{\text{def}}{=} \{\mid\}$ is the only game born on day 0. We can proceed to implement the recursive definition of a game to list the four games born by day 1:

$$0 \stackrel{\text{def}}{=} \{\,|\,\};$$
$$1 \stackrel{\text{def}}{=} \{0\,|\,\};$$
$$-1 \stackrel{\text{def}}{=} \{\,|\,0\};$$
$$* \stackrel{\text{def}}{=} \{0\,|\,0\}.$$

Since these games are so important, we give them their own unique names —
0, 1, −1, and *. The significance of, and justification for, the first three of these
names will become clear. The game * and its relatives are so important that
Chapter 7 will be entirely devoted to their study.

It's important to understand the distinction between the games 0 and *. In
$0 = \{\,|\,\}$ neither player has a legal move; the sets of left and right options are
empty. This is not the same as $* = \{0\,|\,0\}$, where either player has an option
to move to 0 (and so the two games cannot have the same birthday).

The games born by day 2 must have all of their left and right options born
by day 1. Since there are 16 possible subsets of the set of the four games born
by day 1, there appear to be 256 games born on day 2. In fact, as we will
see in Chapter 6, many of these 256 games turn out to be equal in the sense
suggested by the definition given in (4.4), and there are only 22 distinct games
born by day 2.

Conway observed that you can define games with transfinite birthdays.
For example, one can play NIM with an infinite ordinal number of counters in
a heap. We have seen one version in POLYNOMIAL NIM.[2] We will barely touch
on such transfinite games in this book, but the curious reader who wants more
than a taste should refer to *ONAG* [Con01] or [Knu74].

If G is allowed to be an element of $\mathcal{G}^L$ or $\mathcal{G}^R$ (or, more generally, if G
appears anywhere in its own game graph), we arrive at *loopy* games; loopy
games are discussed in detail in *ONAG* [Con01] and *WW* [BCG01]. More
recently, Aaron Siegel has obtained significant results in this area [Sie05].

For this text, the reader should assume that each game has a finite birthday
and so each game has a finite number of options. The following lemma and
corollary justify the "and so" in the last sentence:

Lemma 4.2. *The number of games born by day n, call it $g(n)$, is finite.*

Proof: Since a game born by day n is given by a pair of sets of games born by
day $n-1$, the number of distinct games born by day n is at most $2^{g(n-1)} \cdot 2^{g(n-1)}$.
(Keep in mind that $\mathcal{G}^L$ and $\mathcal{G}^R$ are sets, not multi-sets, so each distinct option

---

[2]The simplest version of this game allows heaps of undetermined size where the first move
is to simply name a (finite) size for the heap — these correspond to heaps of size ω. An
alternative version is to have counters in place of heaps. These are placed on the grid of
non-negative integers, and a legal move for a counter from position (x, y) is to any position
(a, b) with $a \le x$ and either $a < x$ or $b < y$. Such counters correspond to heaps of size $x \cdot \omega + y$.

can appear only once.) By induction $g(n-1)$ is finite, and so we can conclude $g(n)$ is as well.   □

Corollary 4.3. *If G has a finite birthday, then G must also have a finite number of distinct options.*

Proof: Game G with finite birthday n must have options born by day $n-1$, and there are only $g(n-1)$ options from which to choose. So, $\mathcal{G}^L$ and $\mathcal{G}^R$ are finite.   □

If you are *still* worried about the recursive definition of a game, you are now in a position to replace it by an inductive one. Namely, the "games born on day n" can be defined inductively in terms of "games born before day n" (either using $\{\ |\ \}$ as an explicit base case on day 0 or noting that no base case is actually necessary), and then a "game" is simply a "game born on day n" for some n. However, the recursive, or top-down, view of games is a much more useful one than the inductive, or bottom-up, view when proving theorems or carrying out analysis.

Definition of addition

Conway defines addition by

$$G + H \stackrel{\text{def}}{=} \{\mathcal{G}^L + H,\ G + \mathcal{H}^L \mid \mathcal{G}^R + H,\ G + \mathcal{H}^R\}.$$

This introduces a couple of abuses of notation requiring explanation: G and H are games, while $\mathcal{G}^L$, $\mathcal{G}^R$, $\mathcal{H}^L$, and $\mathcal{H}^R$ are sets of games. So what is meant by $\mathcal{G}^L + H$? We define the addition of a single game, G, to a set of games, $\mathcal{S}$, as the set of games obtained by adding G to each element of $\mathcal{S}$:

$$G + \mathcal{S} = \{G + X\}_{X \in \mathcal{S}}.$$

The other abuse of notation is the use of the comma between two sets of games. This comma is intended to mean set union. This notation turns out to be more intuitive and less cumbersome than, say,

$$G + H = \{(\mathcal{G}^L + H) \cup (G + \mathcal{H}^L) \mid (\mathcal{G}^R + H) \cup (G + \mathcal{H}^R)\},$$

for once we have decided to remove braces around sets, treating them essentially as lists, a comma is the natural way to join two lists.

Let's confirm that this notion of addition matches up with that explained by the DOMINEERING position on page 60. We said that in the sum of two games G and H, a player can move on either summand. Left, for instance, could move on the first summand, moving G to some $G^L \in \mathcal{G}^L$, leaving H

alone. A typical left option from the sum $G + H$ is therefore $G^L + H$, as in the definition.

As an example, here are two positions, shown expanded (as in the definition of a game) by listing their left and right options:

$$G = \{\ \blacksquare \mid \blacksquare\ \}$$

$$H = \{\ \blacksquare \mid \blacksquare, \blacksquare\ \}$$

Now, if we mechanically follow the scheme for adding $G + H$, we get

$$G + H =$$

$$\left\{ \underbrace{\blacksquare + \blacksquare, \blacksquare + \blacksquare}_{\mathcal{G}^L + H} \quad \underbrace{\blacksquare + \blacksquare}_{G + \mathcal{H}^L} \;\middle|\; \underbrace{\blacksquare + \blacksquare, \blacksquare + \blacksquare}_{\mathcal{G}^R + H} \quad \underbrace{\blacksquare + \blacksquare, \blacksquare + \blacksquare}_{G + \mathcal{H}^R} \right\},$$

and this corresponds exactly to what we would expect for the left and right options from the position

Theorem 4.4. $G + 0 = G$.

Proof: $G + 0 = \{\mathcal{G}^L \mid \mathcal{G}^R\} + \{\ \mid\ \} = \{\mathcal{G}^L + 0, G + \emptyset \mid \mathcal{G}^R + 0, G + \emptyset\}$. But, for any set $\mathcal{S}$, $\mathcal{S} + \emptyset = \emptyset$,[3] and by induction, $\mathcal{G}^L + 0 = \mathcal{G}^L$ and $\mathcal{G}^R + 0 = \mathcal{G}^R$, so the last expression simplifies to $\{\mathcal{G}^L \mid \mathcal{G}^R\} = G$. $\qquad\square$

Theorem 4.5. *Addition is commutative and associative. That is,*

(1) $G + H = H + G$, *and*

(2) $(G + H) + J = G + (H + J)$.

Exercise 4.6. Prove, by induction, Theorem 4.5.

Definition of negative

The definition of negative,

$$-G \stackrel{\text{def}}{=} \{-\mathcal{G}^R \mid -\mathcal{G}^L\},$$

corresponds exactly to reversing the roles of the two players. (As you might expect, taking the negative of a set negates all the elements of the set; i.e., $-\mathcal{G}^R = \{-G^R\}_{G^R \in \mathcal{G}^R}$.) We swap the left and right options, and recursively swap the roles in all the options.

---

[3] To see why, a game is in $\mathcal{S} + \emptyset$ if it is of the form $A + B$ for $A \in \mathcal{S}$ and $B \in \emptyset$. But there is no $B \in \emptyset$.

Exercise 4.7. Negate the DOMINEERING position

using the formal definition. Confirm that the resulting game has exactly the same game tree as the DOMINEERING position obtained when the roles of the players are reversed by rotating the position through 90 degrees.

Exercise 4.8. Show that, for any game G, (a) $o_L(-G) = \text{☺}$ if and only if $o_R(G) = \text{☺}$ and (b) $o_L(-G) = \text{☺}$ if and only if $o_R(G) = \text{☺}$.
   Use these to convince yourself that (i) if $o(G) = \mathcal{L}$ then $o(-G) = \mathcal{R}$, (ii) if $o(G) = \mathcal{P}$ then $o(-G) = \mathcal{P}$, and (ii) if $o(G) = \mathcal{N}$ then $o(-G) = \mathcal{N}$.

We can now define
$$G - H = G + (-H).$$

Exercise 4.9. Prove that $-(-G) = G$.

Exercise 4.10. Prove that $-(G + H) = (-G) + (-H)$.

Definition of game equivalence

We defined
$$G = H \quad \text{if} \quad (\forall X)\, o(G + X) = o(H + X).$$
In essence, $G = H$ if $G + X$ has the same outcome as $H + X$; that is, G acts like H in any sum of games.

Exercise 4.11. Confirm that $=$ is an equivalence relation; that is, it is reflexive, symmetric, and transitive.

   On the face of it, this definition appears natural, but not very useful. It captures the notion of "being able to substitute G for H," but in order to show $G = H$ it seems that we need to try out *every* possible context X. In this section, we will develop the machinery for an equivalent definition that is much more useful in practice. As a side effect, we will prove that when $G = H$ you can generally substitute G for H wherever you wish, thereby justifying the use of the symbol $=$.
   We will begin by spelling out in a little more detail the proof of Proposition 3.4 in terms of our newly defined notion of equality.

Theorem 4.12. $G = 0$ *if and only if G is a $\mathcal{P}$-position (i.e., G is a win for the second player).*

Proof:
   $\Rightarrow$ If $G = 0$, then $o(G + X) = o(0 + X) = o(X)$ for all X. In particular, when $X = 0$, this says G has the same outcome as 0, which is a $\mathcal{P}$-position.

⇐ Let G be a $\mathcal{P}$-position. Fix any position X; we wish to show that $o(G + X) = o(X)$.

If $o_R(X) = \odot$ then by Lemma 3.3, the lemma showing how Left merely needs to responds locally, $o_R(G+X) = \odot$. If $o_R(X) = \smiley$ then Right wins by moving to some X^R. He can win moving first on $G + X$ by moving to $G + X^R$ and then responding locally. Therefore, $o_R(X) = o_R(G + X)$.

We can argue symmetrically that $o_L(X) = o_L(G + X)$. Hence, X and $G + X$ have the same outcome. $\qquad\square$

Example 4.13. The following diagram shows part of the game tree for an AMAZONS position with value 0. In particular, the diagram shows winning responses to some typical first moves:

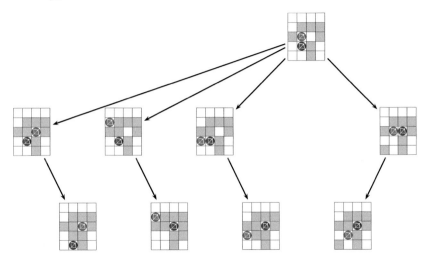

As a consequence, both players can safely ignore that portion of the board when searching for guaranteed winning moves in the following position:

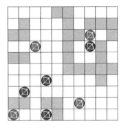

Exercise 4.14. Convince yourself that the original AMAZONS position has value 0. First, confirm that the second player wins in the lines shown. Second, rule out other promising first moves omitted from the game tree.

Corollary 4.15. $G - G = 0$.

Proof: The second player wins on $G-G$ by playing the Tweedledum-Tweedledee strategy. In other words, if the first player moves from $G - G$ to, say, $G - H$, the second player can match the option in the other component, moving to $H - H$, and can proceed to win by induction. $\qquad\square$

Theorem 4.16. *Fix games G, H, and J. Then $G = H$ if and only if $G + J = H + J$.*

Proof:

⇒ Suppose that $G = H$. We wish to show that, for all X, $(G+J)+X$ has the same outcome as $(H + J) + X$. If we choose $X' = (J + X)$ and apply the definition of $G = H$, then $G+X' = H+X'$; i.e., $G+(J+X) = H+(J+X)$ and associativity (Theorem 4.5) yields $(G + J) + X = (H + J) + X$.

⇐ Suppose that $G + J = H + J$. We can use the ⇒ direction just proved to conclude that $(G + J) + (-J) = (H + J) + (-J)$. By associativity and noting that $J - J = 0$, we have

$$G = G + 0$$
$$= G + (J - J)$$
$$= (G + J) - J$$
$$= (H + J) - J$$
$$= H + (J - J)$$
$$= H + 0$$
$$= H. \qquad\square$$

Exercise 4.17. Identify which theorem or definition justifies each $=$ in the last expression.

Exercise 4.18. Prove that if $G = G'$ and $H = H'$, then $G + H = G' + H'$.

Corollary 4.19. $G = H$ *if and only if $G-H = 0$; that is, $G-H$ is a $\mathcal{P}$-position.*

Proof: Simply add $-H$ to both sides and use associativity. $\qquad\square$

This corollary is important, for it gives a clear, constructive way to determine whether $G = H$: simply play the game $G - H$ and see if the second player always wins. In practice, this is the easiest and most common way of testing whether two games are equal. This completes a circle begun on page 66. We've provided a "middle road" definition of equality in order to allow for later gains in simplifying positions to determine their outcome classes. On the other hand, at this point, in order to determine whether or not two games G and H are equal, we must return to the pragmatic position and answer the question of whether or not $G - H \in \mathcal{P}$. However, once we have this information, it allows us to replace G by H freely in any context — which, if say H is much simpler than G, will be of tremendous value in later analysis. Furthermore, in Section 4.3, we will provide *automatic* means for finding a unique simplest form for any game G, sparing us the guesswork of trying to find suitably simple games H for which $G = H$.

Definition of greater than or equal

We state that

$$G \geq H \text{ if } (\forall X) \, o(G + X) \geq o(H + X),$$
$$G \leq H \text{ if } (\forall X) \, o(G + X) \leq o(H + X).$$

In short, this says that $G \geq H$ if replacing H by G can *never* hurt Left, no matter what the context, assuming that she plays optimally.

Exercise 4.20. Confirm that $G \geq H$ if and only if $H \leq G$. Also confirm that $G = H$ if and only if $G \geq H$ and $G \leq H$.

As with our original definition of $G = H$, while this definition is reasonably intuitive, it is not very constructive. In this section, we will give a more constructive definition for how to test if $G \geq 0$ and generalize that notion to test if $G \geq H$.

Theorem 4.21. *The following are equivalent:*

(1) $G \geq 0$.

(2) *Left wins moving second on G; that is, $G \in \mathcal{P}$ or $G \in \mathcal{L}$.*

(3) *For all games X, $o_R(G + X) \geq o_R(X)$.*

(4) *For all games X, $o_L(G + X) \geq o_L(X)$.*

Proof:

$1 \Leftrightarrow$ both 3 and 4 hold: The outcome class of a game is determined by whether Left wins moving first and whether Left wins moving second.

$2 \Rightarrow 3$: This is Proposition 3.5.

$2 \Rightarrow 4$: The proof is analogous to $(2 \Rightarrow 3)$.

$3 \Rightarrow 2$: Choose $X = 0$, which gives $o_R(G + 0) \geq o_R(0) = \smiley$.

$4 \Rightarrow 2$: Since $o_L(G + X) \geq o_L(X)$ for all games X, then let $X = -G$. This gives $o_L(G - G) = o_L(0) = \odot \geq o_L(-G)$, that is, $o_L(-G) = \odot$. From Exercise 4.8 we then have that $o_R(G) = \odot$. $\qquad\square$

Exercise 4.22. Prove $2 \Rightarrow 4$ in Theorem 4.21.

Theorem 4.23. $G \geq H$ *if and only if* $G + J \geq H + J$, *for all games* G, H, *and* J.

Proof: The proof parallels that of Theorem 4.16. $\qquad\square$

Theorem 4.24. $G \geq H$ *if and only if Left wins moving second on* $G - H$.

Proof: $G \geq H$ if and only if $G - H \geq H - H = 0$ (i.e., $G - H \geq 0$). $\qquad\square$

This last theorem is how, in practice, one should compare G with H. In particular, when comparing G with H, determine the outcome of $G - H$. If $G - H$ is a $\mathcal{P}$-position, then $G = H$. If $G - H$ is an $\mathcal{L}$-position, then $G \geq H$ but $G \neq H$, that is, $G > H$. If $G - H \in \mathcal{R}$, we have $G < H$. Lastly, if $G - H$ is in $\mathcal{N}$, then it is neither the case that $G \geq H$ nor that $G \leq H$; G and H are incomparable! When two games are incomparable, we say that G is *confused with* or *incomparable* with H and denote this by $G \parallel H$.

To recap,

$G > 0$ when L wins G	$G > H$ when L wins $G - H$
$G = 0$ when 2nd wins G	$G = H$ when 2nd wins $G - H$
$G < 0$ when R wins G	$G < H$ when R wins $G - H$
$G \parallel 0$ when 1st wins G	$G \parallel H$ when 1st wins $G - H$

Lastly, just as we write $G \geq H$ to mean $G = H$ or $G > H$, we will use the symbol $G \rhd H$ to mean $G > H$ or $G \parallel H$; that is, G is greater than or incomparable to H. Note that $G \rhd H$ is equivalent to $G \not\leq H$, but is somewhat more intuitive. Similarly, $G \lhd H$ means that $G < H$ or $G \parallel H$.

4.2   Games Form a Group with a Partial Order

It is easy to prove that $\geq$ is a partial order as long as we use game-equality (which we have written as $=$) for equality. To be a partial order, a relation must be transitive, reflexive, and antisymmetric:

Transitive: If $G \geq H$ and $H \geq J$, then $G \geq J$.

Reflexive: For all games G, $G \geq G$.

Antisymmetric: If $G \geq H$ and $H \geq G$, then $G = H$.

Theorem 4.25. *The relation $\geq$ is a partial order on games.*

Proof:

Transitive: Suppose that $G \geq H$ and $H \geq J$. Then, Left wins moving second on $(G - H)$ and on $(H - J)$, and so by responding locally (Lemma 3.3), she can win moving second on the sum $(G - H) + (H - J)$. However, $(G - H) + (H - J) = (G - J) + (H - H)$, which has the same outcome as $(G - J)$. Hence, $G \geq J$.

Reflexive: Left wins moving second on $G - G = 0$.

Antisymmetric: Given by Exercise 4.20.   □

Games also form an *abelian (or commutative) group*. A *group* is a collection of elements (games) along with a binary operation (here, $G + H$), an inverse $(-G)$, and an identity (the game 0), which satisfy the following:

Closure: If G and H are elements of the group, so is $G + H$.

Associativity: For all games G, H, and J, we have $(G+H)+J = G+(H+J)$.

Identity: For all games G, $G + 0 = G$.

Inverse: For all games G, $G + (-G) = 0$.

The group is *abelian* if, in addition, the operation is commutative; that is, $G + H = H + G$.

The main subtlety here is that we are really talking about the group structure of games under the equivalence relation $=$. In particular, there is not just one zero game $0 = \{ \,|\, \}$, but rather any game equal to 0 (i.e., any $\mathcal{P}$-position) is an identity.

Theorem 4.26. *Games form an abelian group.*

Proof:

Closure: $G + H \stackrel{\text{def}}{=} \{G^L + H,\ G + \mathcal{H}^L \,|\, G^R + H,\ G + \mathcal{H}^R\}$ is a pair of sets of games (by induction), and so $G + H$ is a game.

Associativity and commutativity: Theorem 4.5.

Identity: Theorems 4.4 and 4.12.

Inverse: Corollary 4.15.   □

Example 4.27. What are the order relationships between the four games born by day 1 (i.e., 0, 1, -1, and $*$)? Which sums (if any) of games born by day 1 are also born by day 1?

We already know that $0 \in \mathcal{P}$, $1 \in \mathcal{L}$, $-1 \in \mathcal{R}$, and $* \in \mathcal{N}$. Since $G + 0 = G$ for any game G, it follows that $-1 < 0$, $0 < 1$, and $0 \parallel *$. We could, but don't need to, check that $-1 < 1$ since this follows by transitivity. What about the order relationships between -1 and $*$ or between 1 and $*$? First, note that $* + * = 0$. (We will generalise this below, but it is clear that whoever moves first in $* + *$ changes one of the $*$s to 0, and the second player then wins by playing in the other $*$.) So, $* = -*$. Now, to determine the order relationship between $*$ and 1, we need to know the outcome class of $1 - *$, or equivalently of $1 + *$. But,

$$o_L(1 + *) = \smiley \quad \text{since Left can move to 1 and Right then has no move,}$$

$$o_R(1 + *) = \smiley \quad \text{since Right's only move is to 1 and Left then wins.}$$

Thus, $o(1+*) = \mathcal{L}$, and therefore $1 > *$. Similarly, $* > -1$ (or we could cleverly note that $* > -1$ if and only if $1 + * > 1 + (-1) = 0$, which is what we just proved!).

As regards sums we can also make use of things that we have already proven. Certainly $0 + X = X$ for any X, so all the sums of 0 with a game born by day 1 are also born by day 1. Furthermore, we know that $* + * = 0$ and $1 + (-1) = 0$, so these are also born by day 1. The remaining games to consider are $1 + 1$, $1 + *$, $-1 + *$, and $(-1) + (-1)$. By symmetry (or taking negatives), we only need to worry about the first two. Since $1 > 0$, $1 + 1 > 1$, but we know that 1 is the largest game born by day 1, so $1 + 1$ is not born by day 1. Also since $*$ 0, $1 + *$ 1, and every game born by day 1 is comparable to 1, so $1 + *$ is not born by day 1 either.

We now generalize an observation of the previous example. Recall that an impartial game is one all of whose followers have the same Left and Right options (i.e., the moves available in any position are the same regardless of whose turn it is).

Proposition 4.28. *Let G and H be impartial games. Then, $G = -G$ and either $G = H$ or G H.*

Proof: Since G is impartial, either player can play the Tweedledum-Tweedledee strategy in $G + G$. So $G + G \in \mathcal{P}$, i.e., $G + G = 0$, i.e., $G = -G$. If G and H are impartial and $G \neq H$, then what can the outcome class of $G - H$ be? Since $G - H = G + H$ is impartial, its outcome class must be $\mathcal{P}$ or $\mathcal{N}$ (it can't be $\mathcal{L}$ since Right could just steal Left's strategy as first player — and so, of course, it can't be $\mathcal{R}$ either). But if $G \neq H$ then $G + H = G - H \neq 0$, so $G - H \in \mathcal{N}$ and G H. $\qquad\square$

Exercise 4.29. For each of the following six separate positions — two TOP-PLING DOMINOES, two MAIZE (shown with ⊙), and two MAZE (⍟) — determine the outcome class for each position and whether it equals a game born by day 1 (either 0, 1, −1, or ∗).

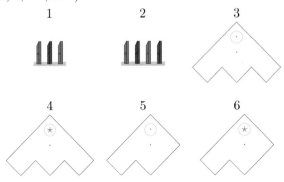

Example 4.30. Show that in DOMINEERING,

Proof: To show

it suffices to show that Left wins

whether she moves first or second. All the relevant lines of play are summarized below:

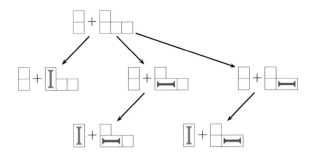

In particular, if she moves first, she wins by moving on the second component to

$$\boxed{} + \boxed{\text{I}} = 0.$$

Moving second, Right can only make one move no matter what Left does and is doomed. □

Example 4.31. Show that, in CLOBBER, ▨▨▨▨ > ▨▨ but that ▨▨▨▨ is confused with ▨▨▨.

First, consider the difference game ▨▨▨▨ − ▨▨ = ▨▨▨▨ + ▨▨. Left wins moving first or second by playing in the second component (if possible) at her first opportunity, and then finishing the game by taking Right's piece in the first component. Hence, ▨▨▨▨ > ▨▨. Left's winning lines are shown below:

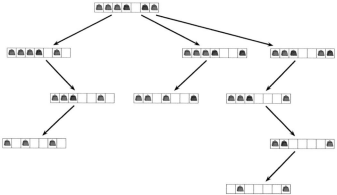

Now consider ▨▨▨▨ − ▨▨▨ = ▨▨▨▨ + ▨▨▨. Left moving first plays to ▨▨▨▨ + ▨▨, and we just showed this wins. If Right moves first, however, he can play to ▨▨▨ + ▨▨▨ = 0 and wins as well. Hence, ▨▨▨▨ ‖ ▨▨▨:

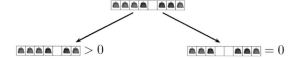

Exercise 4.32.

(a) In DOMINEERING, show that

 > 0

(b) Show that the DOMINEERING position

     is equal to the HACKENBUSH position

4.3   Canonical Form

As we have seen, two games can be equal despite having very different game trees. For example, careful analysis will show that

The surprising, but extremely useful, fact is that every game G has a unique *smallest* game, in size and birthday, which is equal to it. This game is called G's *canonical form,* and there is a clear, methodical procedure to simplify G to reach its canonical form.[4] So, one way to check if G and H are equal is to see whether their canonical forms are identical.

There are two mechanisms for simplifying a game given in the next two subsections. What is remarkable is that these are the only two simplifications that one has to perform to reduce a game to its unique canonical form!

Dominated options

Consider the DOMINEERING position

Left, who plays the vertical dominoes, has two options:

 and

The first of these leaves a position of value 0 since neither player has a move available. The second leaves a position in which Right, but not Left, has a move, a position of value -1. Unless Left is playing the game against a small child who shares a significant portion of her DNA, there are no circumstances when she should choose to play the second move. We say it is *dominated* by the first move, since she would always prefer the situation arising after her first move (whatever the remaining context might be) to that arising from her second move. This is really a version of the One-Hand-Tied Principle — Left promises not to make the second move, knowing that it can never be in her interests to do so. We now generalize and formalize this idea.

Theorem 4.33. *If*
$$G = \{A, B, C, \ldots \mid H, I, J, \ldots\}$$
and $B \geq A$, then $G = G'$ where
$$G' = \{B, C, \ldots \mid H, I, J, \ldots\}.$$

---

[4]This is a procedure that gives a unique representative of the equivalence class defined by $=$.

Similarly for Right, if $I \geq H$, then option I can be removed. Here, option B is said to *dominate* option A for Left and H dominates I for Right, and the theorem allows us to *remove a dominated option*.

Proof: To see that $G = G'$, we will confirm that $G - G' = 0$; that is, that either player wins moving second on $G - G'$. On this difference game, if Left moves G to B or C, Right responds by moving $-G'$ to $-B$ or $-C$, respectively. In fact, all moves by Left or Right on G pair up with a move by the opponent on $-G'$, except the move to A. The second player can respond to any first-player move on $-G'$ by making the corresponding move on G. So, the only way the first player might hope to avoid a simple Tweedledum-Tweedledee strategy is to play Left and move G to A. But then, Right responds by moving G' to $-B$, leaving $A - B$, which, since $B \geq A$, Right wins as the second player.

This argument is summarized in the following diagram. We have paired up moves with obvious responses with $\sqrt{}$s. For example,

$$\text{if Left moves to } \overset{\checkmark}{B},$$

$$\text{Right responds on the other component to } -\overset{\checkmark}{B}$$

and vice versa:

Since Right has an advantage, the diagram shows:

$$\underbrace{\left\{ \overset{\checkmark}{A}, \overset{\checkmark}{B}, \overset{}{C}, \ldots \mid \overset{\checkmark}{H}, \overset{\checkmark}{I}, \overset{\checkmark}{J}, \ldots \right\}}_{G} - \underbrace{\left\{ \overset{\checkmark}{B}, \overset{\checkmark}{C}, \ldots \mid \overset{\checkmark}{H}, \overset{\checkmark}{I}, \overset{\checkmark}{J}, \ldots \right\}}_{G'}$$

$$A - G'$$

$$A - B$$
$$\text{R wins}$$
$$\text{since}$$
$$B \geq A$$

□

Reversible options

Using "reversibility" can be confusing, so let's set a strategic situation. In a game G, Left is in a bad way since Right has an advantage. Left makes a move that threatens this advantage and Right "reverses" Left's move to regain the advantage. Since Right will do this automatically, Left really doesn't have the original option but is faced with options after Right's reversing move. "Reversibility" is a more general concept but with the same idea: Right's situation is at least as good after his move compared to his situation in the original game. The reader needs to be aware that the full power of reversing moves only occurs in more complicated games than we've presented so far.

To get the second procedure that will lead to the canonical form, we need to formalize. Let A be a left option of a game G, and suppose that there is some $A^R \leq G$. We say that A is a *reversible option,* reversible through A^R (or A^R is the *reversing option*) and A^{RL} is the *replacement set.*

That is, a left option A of G can be considered to be reversible if A has a right option A^R that, from Right's point of view, is just as good or better than G; that is, $A^R \leq G$. In any context containing G, Right can make a One-Hand-Tied promise, "if you ever choose option A of G, then I will immediately move to A^R." When Right makes such a promise, it doesn't make sense for Left to choose option A *unless* she intends to follow up Right's move to A^R with an immediate response to one of the left options of A^R. If she plans some other move elsewhere, she might just as well start with that. So, Left can consider the left options of A^R as being *immediately* available to her from G instead of worrying about the interchange "first I choose A, then Right chooses A^R, then I choose one of its left options."

To clarify the "explanation," particularly when the replacement set is empty,[5] let's consider the games $G = \{\{2 \mid \{-8 \mid -9\}\} \mid -1\}$ and $H = \{1 \mid \{9 \mid 2\}\}$. First consider G. In $G - \{-8 \mid -9\} = G + \{9 \mid 8\}$, Right cannot win playing first and Left can, so $G - \{-8 \mid -9\} > 0$ and $G > \{-8 \mid -9\}$. Therefore, $\{2 \mid \{-8 \mid -9\}\}$ is a reversible option, $\{-8 \mid -9\}$ is the reversing option, and $\{-8\}$ is the replacement set for $\{2 \mid \{-8 \mid -9\}\}$. If this approach works (and it does), then $G = \{-8 \mid -1\}$. Now let's look at H. The reader can check that $9 - H \geq 0$, which gives $9 > H$. This means that $\{9 \mid 2\}$ is a reversible option, and 9 is the reversing option. Since the game 9 means that Left has 9 moves and Right none, the replacement set is empty. Informally, this is called *reversing out* because the reversible option is replaced by nothing. Here, H becomes $\{2 \mid \}$, which we know as the game 3.

Reversibility is our second simplification principle that leads to canonical forms for games.

This simplification rule is not at all intuitive and may take some getting used to. To repeat:

Theorem 4.34. *Fix a game*

$$G = \{A, B, C, \ldots \mid H, I, J, \ldots\}$$

and suppose that for some right option of A, call it A^R, $G \geq A^R$. If we denote

---

[5]Reversible options occur in misère games, scoring games, and short disjunctive sum games, but the situation is complicated if the replacement set is the empty set.

the left options of A^R by $\{W, X, Y, \ldots\}$,

$$A^R = \{W, X, Y, \ldots \mid \ldots\},$$

and define the new game

$$G' = \{W, X, Y, \ldots, B, C, \ldots \mid H, I, J, \ldots\},$$

then $G = G'$.

When the theorem applies, we have a reversible option A, a reversing option A^R, and a set of replacements for A. It is possible that there are no left options of A^R so there are no replacements, in which case A is eliminated as if it were a dominated option. In this case, the move to A is said to *reverse out* through A^R. The process of replacing A with the left options of A^R is called *bypassing the reversible option A.*

Proof: Assume all the hypotheses of the theorem. To show that $G = G'$, we will show that the second player wins the difference game $G - G'$. First we summarize the proof with a diagram:

The heart of the proof lies in the two assertions justified in the words "... can win since ...". To address the first, to find a winning move $A^R - H$, Right uses the fact that $A^R \leq G$ (by the definition of reversible option). In particular, Right wins moving second on $A^R - G$, and so Right has a winning response to *every* Left option from $A^R - G$, and $A^R - H$ is one such left option.   □

Exercise 4.35. Summarize the diagrammed proof in words. As you do so, convince yourself that every case is handled. Be sure to understand the second case in which the diagram has the words, "... can win since ...".

We will see examples of how to simplify a game by bypassing reversible options in the next chapter. We are now ready to prove the main theorem of the section.

Reduction to canonical form

We say that G is in *canonical form* if G and all of G's positions have no dominated or reversible options. The following theorem justifies the term *canonical*:

Theorem 4.36. *If G and H are in canonical form and $G = H$, then $G \cong H$.*

In other words, if you start with a game G, bypass all its reversible options, and remove all its dominated options until no more simplification is possible, then you arrive at the unique smallest game, called G's canonical form, which is equivalent to G. There are two important things to note about this process. First, both the removal of a dominated option and the reversal of a reversible option produce a game whose game tree has fewer nodes than the original tree did. Therefore, the process of removing dominated options and bypassing reversible ones must terminate. Second, the order in which we choose to carry out this procedure turns out to be irrelevant — there may well be choices, but they will all lead to the same canonical form. The theorem also shows that no other reduction process is required.

This important theorem is surprisingly easy to prove.

Proof: Suppose that G, H, and all their positions have no reversible or dominated options. By induction, their left and right options are in canonical form, and it suffices to show that G's left options match up with H's.

Since $G = H$, Left wins moving second on $G - H$. In particular, Left has a winning response to $G^R - H$. That winning response cannot be in G^R, for then we would have that some $G^{RL} - H \geq 0$, and so $G^{RL} \geq H = G$, and so G would have a reversible option. Thus, the winning response to $G^R - H$ must be on the second component to some $G^R - H^R$, and we have that $G^R \geq H^R$ for some H^R.

We can construct the same argument for any initial move on G or H. In particular, for each G^R there exists an H^R such that $G^R \geq H^R$. A parallel argument proves that for each H^R there exists a $G^{R'}$ such that $H^R \geq G^{R'}$. Since $G^R \geq H^R \geq G^{R'}$, G^R and $G^{R'}$ must be identical (for otherwise G^R is dominated). So, every right option of G equals some right option of H; i.e., $\mathcal{G}^R \subseteq \mathcal{H}^R$. By symmetric arguments, $\mathcal{H}^R \subseteq \mathcal{G}^R$, so $\mathcal{G}^R = \mathcal{H}^R$, and similarly $\mathcal{G}^L = \mathcal{H}^L$. Hence, $G \cong H$.   □

Exercise 4.37. Convince yourself that there isn't a third reduction that, starting with a game G, could generate an equal game of lesser birthday than that of the canonical form of G.

We close with a useful related lemma, which we state here since the proof is similar to that of the last theorem.

Lemma 4.38. *If $G = H$ and G is in canonical form, then each option of G is dominated by an option of H; i.e.,*

$$(\forall G^L)(\exists H^L) such\ that\ H^L \geq G^L,\ and$$
$$(\forall G^R)(\exists H^R) such\ that\ H^R \leq G^R.$$

Proof: The proof is similar to, but simpler than, that of Theorem 4.36, and we leave it as an exercise.   □

Exercise 4.39. Prove Lemma 4.38. (Note that H need not be canonical.)

Exercise 4.40. Suppose that $G = 0$ with $G^{\mathcal{L}} \neq \emptyset$, that is, G is not in canonical form. Show that if $G^L \in G^{\mathcal{L}}$ then G^L is reversible.

Example 4.41. The 4×1 strip for DOMINEERING has the form $F = \{1, 0 \mid \cdot\}$. Eliminating dominated options gives $\{1 \mid \cdot\} = 2$.

The reduction to canonical form can be involved. Given some G, if you have an inspired guess that $G = H$, then the quick test is to verify whether or not $G - H = 0$. In the following examples, there is no guessing, inspired or otherwise.

Example 4.42. The canonical form of $G = \{\{4 \mid 1\}, 0 \mid 2\}$ is $\{0 \mid \} = 1$.

The game $\{4 \mid 1\}$ is a Left win, i.e., $\{4 \mid 1\} > 0$, so 0 is dominated and therefore $G = \{\{4 \mid 1\} \mid 2\}$. Left wins $G - 1$ playing second (if Right plays to $\{\{4 \mid 1\} \mid 2\} - 0 = \{\{4 \mid 1\} \mid 2\}$, Left plays to $\{4 \mid 1\}$ and wins; if Right plays to $2 - 1$, he loses); so $1 = \{0 \mid \}$ is a reversing option for Right, $\{4 \mid 1\}$ is a reversible option, and the corresponding replacement set is $\{0\}$. Therefore, $G = \{0 \mid 2\} = \{0 \mid \{1 \mid \}\}$. Since $G \leq 1$, 2 is a reversible option with 1 the reversing option. Since $1 = \{0 \mid \}$, the replacement set is empty, giving $G = \{0 \mid \}$.

Example 4.43. The canonical form of $G = \{-5 \mid -2\}$ is $G = -3$.

To show this, note that Left has no move from -2 so it is not reversible. From -5, Right has an option to -4, and since $G + 4 \in \mathcal{L}$ then $G > -4$. Therefore, -5 is a reversible option and -4 is the reversing option. Since the replacement set (Left's options from -4) is empty, $G = \{\,\mid -2\} = -3$.

Example 4.44. The game $\{\{4 \mid 2\}, 3 \mid -1\}$ is in canonical form.

The games $\{4 \mid 2\}$ and 3 are incomparable since $\{4 \mid 2\} - 3$ is in $\mathcal{N}$, therefore there are no dominated options. The game 3 is not reversible because it has no Right option. Since $\{\{4 \mid 2\}, 3 \mid -1\} - 2$ is in $\mathcal{N}$, $\{4 \mid 2\}$ is not reversible.

Example 4.45. The canonical form of $G = \{\{3 \mid 1\}, 2 \mid 0\}$ is $\{2 \mid 0\}$.

The games $\{3 \mid 1\}$ and 2 are incomparable ($\{3 \mid 1\} - 2$ is a second-player win), thus there is no domination. Since Left wins $G - 1$ going first and second, $G > 1$. Therefore, $\{3 \mid 1\}$ is a reversible option, 1 is a reversing option for Right, and the replacement set, Left's option from 1, is $\{0\}$. Bypassing the reversible option gives $G = \{0, 2 \mid 3\} = \{2 \mid 3\}$ since 2 dominates 0. Note that $3(= \{2 \mid \})$ is not reversible because $G - 2$ is a first-player win.

Example 4.46. The canonical form of $\{2, \{20 \mid -10\} \mid 1\}$ is $\{2 \mid 1\}$.

The game is positive and the Right option of $\{20 \mid -10\}$ is negative, thus it reverses out and is replaced by the Left option of -10, which does not exist and so $\{20 \mid -10\}$ disappears.

The proof of this example can be generalized to prove the following:

Lemma 4.47. *Let* $G = \{G^{\mathcal{L}} \mid G^{\mathcal{R}}\}$ *with all* $G^{\mathcal{R}} \rhd 0$ *and there exist at least one* $K \geq 0$, $K \in G^{\mathcal{L}}$. *Then, any Left option of* G *that has a negative Right option can be deleted.*

Proof: First, $G > 0$ since Left wins going first or second. Suppose that $H \in G^{\mathcal{L}}$ has a negative Right option, H^{R}. If H^{R} has no Left option, then H can be deleted from $G^{\mathcal{L}}$. If H^{R} has a Left option, then H is replaced by H^{RL}. Each game in H^{RL} is in $\mathcal{R}$ or $\mathcal{N}$. Any option in $\mathcal{R}$ is dominated by K and so is deleted. Any option $H^{RL} \in \mathcal{N}$ must have a Right option in $\mathcal{P}$ or $\mathcal{R}$ and consequently $G > H^{RL}$. Hence, H^{RL} is reversible. Repeat this process, which must finish because the game tree of G is of finite depth. It finishes when all the generated Left options are in $\mathcal{R}$ and are then deleted. That is, H has been deleted.   □

The next example illustrates an important point. Theorem 4.34 only refers to one reversing option; consequently, even if there are many, *only one reversing option need be considered.*

Example 4.48. Let $G = \{2, \{3 \mid 0, *\} \mid 4\}$, then the canonical form of G is 3. Recall that $* = \{0 \mid 0\}$.

Since $\{3 \mid 0, *\} - 2$ is a first-player win (Left moves to $3 - 2 > 0$ and Right moves to $0 - 2 < 0$), the two left options are confused with each other and so there are no dominated options.

Both $G > 0$ and $G > *$ are true, and therefore $\{3 \mid 0, *\}$ is a reversible option with two reversing options.

1. Suppose that 0 is taken as the reversing option; then, the replacement set for $\{3 \mid 0, *\}$ is empty and we can set $G = \{2 \mid 4\}$.

2. Suppose that $*$ is taken as the reversing option; then, the replacement set for $\{3 \mid 0, *\}$ is 0 and we can set $G = \{2, 0 \mid 4\}$. Since $2 > 0$ then $G = \{0 \mid 4\}$.

Finally, consider the right options: $4 = \{3 \mid \}$ has the left option 3; Right wins $\{0 \mid 4\} - 3$ (Left moves to $0 - 3 < 0$, Right moves to $\{0 \mid 4\} - 2$, and Left now has to move to $0 - 2 < 0$), therefore $\{0 \mid 4\} < 3$ and 3 is a reversing option for the reversible option 4. The replacement set consists of the right options from 3, but there aren't any, so $G = \{2 \mid \} = 3$.

Exercise 4.49. Find the canonical form of $\{3, \{4 \mid \{1, \{2 \mid 0\} \mid 0\}\} \mid 3\}$.

4.4   Case Study: Cricket Pitch

The difference between rolling a lawn (amateur activity) and preparing a cricket pitch (professional activity) is profound, to those who like the sport of cricket. Once a perfect pitch is achieved, it should not be tampered with! The rules to CRICKET PITCH are identical to ROLL THE LAWN (from Section 2.4) *except* that the roller is never allowed to roll over a 0. While we know the outcome classes of single CRICKET PITCH positions, we still do not know how to play sums of positions [NO11, Sie11]. We are, however, prepared to analyze the outcomes of single-roller positions using reduction techniques that exploit equality of games.

Reductions for cricket pitch

Let α, β and γ be (possibly empty) strings of non-negative integers. Let $\alpha + 2$ be the string α with every number increased by 2.

Lemma 4.50. *The following two reductions simplify* CRICKET PITCH *positions maintaining equality:*

- *Prune:* $[\alpha, 0, \beta, \odot, \gamma] = [\beta, \odot, \gamma]$;

- *Reduce:* $[\alpha, \odot, \beta] = [\alpha + 2, \odot, \beta + 2]$.

Proof: Since the roller cannot go past a 0, then it is clear that $[\alpha, 0, \beta, \odot, \gamma] - [\beta, \odot, \gamma]$ is a second-player win because the first player's moves can be copied by the second player in the other component.

For the second reduction, it suffices to show that

$$[\alpha + 2, \odot, \beta + 2] - [\alpha, \odot, \beta]$$

is a second-player win (i.e., is 0). Further, by symmetry, we need only consider one player moving first (say, Left). Recall that roles are reversed on the second (negative) summand.

If Left moves in the second summand (where she moves right!) to

$$[\alpha_1 + 2, \odot, \beta + 2] - [\alpha, \beta_1, \odot, \beta_2],$$

then Right responds with the "mirror" move in the first summand to

$$[\alpha_1 + 2, \beta_1 + 2, \odot, \beta_2 + 2] - [\alpha, \beta_1, \odot, \beta_2],$$

which is 0 by induction. If Left moves to

$$[(\alpha_1 + 2), \odot, \alpha_2 + 1, \beta + 2] - [\alpha, \odot, \beta],$$

then, if possible, Right mirrors by playing to

$$[\alpha_1 + 2, \odot, \alpha_2 + 1, \beta + 2] - [\alpha_1, \odot, \alpha_2 - 1, \beta],$$

which is zero by induction. If this move is not possible, then we know that Left moved over a 2 in $\alpha + 2$. That is, $\alpha = \delta, 0, \gamma$, where γ does not contain any 0s. In this case, Left initially moved from $[\delta + 2, 2, \gamma + 2, \odot, \beta + 2] - [\delta, 0, \gamma, \odot, \beta]$ to $[\delta_1 + 2, \odot, \delta_2 + 1, 1, \gamma + 1, \beta + 2] - [\delta, 0, \gamma, \odot, \beta]$. We claim that Right wins by moving to

$$[\delta_1 + 2, \odot, \delta_2 + 1, 1, \gamma + 1, \beta + 2] - [\delta, 0, \odot, \gamma - 1, \beta].$$

From here, if Left moves in the second component, then Right mirrors Left's move in the other component and the resulting position is 0 by induction. Otherwise, Left moves in $[\delta_1 + 2, \odot, \delta_2 + 1, 1, \gamma + 1, \beta + 2]$, and then Right responds in the same component moving over but not farther than the bump of size 1, i.e., ignoring the bumps that are no longer reachable. He moves to

$$[0, \odot, \gamma + 1, \beta + 2] - [0, \odot, \gamma - 1, \beta],$$

and this is 0 by induction. □

Outcomes of cricket pitch

The basic observation is that even numbers contribute little since they will get passed over an even number of times. In ROLL THE LAWN this is the only required observation. The only case in which an even number would not take part in an even number of moves is if there is a smaller odd number on the opposite side from the roller.

Definition 4.51. Let G be the CRICKET PITCH position $[\alpha, \odot, \beta]$, $\alpha = a_1, a_2, \ldots, a_n$ and $\beta = b_m, b_{m-1}, \ldots, b_1$. The *Left odd low point*, $\mathrm{ldip}(G)$, is $\mathrm{ldip}(G) = \min\{a_i : a_i$ is odd and $a_i < a_j, i < j\}$. If there is no such bump, then $\mathrm{ldip}(G) = \infty$. Similarly, $\mathrm{rdip}(G) = \min\{b_i : b_i$ is odd and $b_i < b_j, i < j\}$.

For example, in $[1, 2, 3, 5, \odot, 1, 2, 3]$, the bumps a_1, a_3 and a_4 are odd and smaller than the bumps between themselves and the roller, but $a_1 = 1$ is the least, and $\mathrm{ldip}(G) = 1$. Note that $\mathrm{rdip}(G) = 1$ since the bump immediately to the right of the roller is odd and there is no smaller bump between it and the roller. In $[3, 2, 3, 4, \odot, 2, 2, 2, 3]$, $\mathrm{ldip}(G) = 3$ and $\mathrm{rdip}(G) = \infty$.

Theorem 4.52. *Let G be a CRICKET PITCH position. The outcome classes are determined by the odd low points:*

(1) If $\mathrm{ldip}(G) < \mathrm{rdip}(G)$, then $o(G) = \mathcal{L}$.

(2) If $\mathrm{ldip}(G) > \mathrm{rdip}(G)$, then $o(G) = \mathcal{R}$.

(3) If $\mathrm{ldip}(G) = \mathrm{rdip}(G) < \infty$, then $o(G) = \mathcal{N}$.

(4) If $\mathrm{ldip}(G) = \mathrm{rdip}(G) = \infty$, then $o(G) = \mathcal{P}$.

Proof: Suppose that $\mathrm{ldip}(G) = \mathrm{rdip}(G) = \infty$. In this case, either all the bumps are even or any odd-sized bump has at least one smaller even-sized bump between it and the roller. In the process of reducing and pruning, any odd bump will get pruned and all the even bumps will be reduced to 0 or will be pruned. Therefore, the final reduced-and-pruned position is $\odot$ with no bumps and $o(G) = \mathcal{P}$.

In all other cases, let G' be the position after all possible reducing and pruning is done. In G', there is a bump of size 1 that we may assume is on the left of the roller so that $\mathrm{ldip}(G') = 1$.

For part (1), $\mathrm{rdip}(G') > 1$. Now Left can win G', and so G, playing first or second since she can always move past the 1, leaving Right with no move. A similar argument holds for part (2). For part (3), G' will have $\mathrm{ldip}(G') = \mathrm{rdip}(G') = 1$, and now both players have a winning first move. $\qquad\square$

Example 4.53. Let $G = [6, 9, 5, 6, 4, 9, 6, 7, 7, 7, 9, 11, ⊙, 13, 11, 7, 4, 6, 6, 9, 8, 8]$. Since $\mathrm{ldip}(G) = 7$ and $\mathrm{rdip}(G) = 7$, G is a next-player win. Apply the following reductions:

$$\text{reduce} — [2, 5, 1, 2, 0, 5, 2, 3, 3, 3, 5, 7, ⊙, 9, 7, 3, 0, 2, 2, 5, 4, 4],$$
$$\text{prune} — [5, 2, 3, 3, 3, 5, 7, ⊙, 9, 7, 3],$$
$$\text{reduce} — [3, 0, 1, 1, 1, 3, 5, ⊙, 7, 5, 1],$$
$$\text{prune} — [1, 1, 1, 3, 5, ⊙, 7, 5, 1].$$

The reductions show that Left can move to a position equivalent to $[1, 1, ⊙, 0]$ and Right to $[0, ⊙] = 0$.

4.5 Incentives

It is often helpful, when playing games, to focus on the amount gained or lost by a move rather than on the resulting position.

Definition 4.54. The *left incentives* and *right incentives* of a game are the sets $G^L - G$ and $G - G^R$, respectively. The *incentives* of a game are the union of the left and right incentives. Similarly, the left incentive (or, respectively, right incentive) of a particular move G^L is $G^L - G$ (or $G - G^R$).

Note, as you compare the definitions of left and right incentives, that the order of terms in the difference has switched. Consequently, both players should favor moves of higher incentive. This makes it meaningful, for example, to compare the left incentives of G against its right incentives. In particular, when playing in a sum of games $A + B + C + D + E + F + G$, Left should choose to play in the component that has largest left incentive (if such a component exists). The incentives are games and all the theory of canonical forms can be used.

Example 4.55.

(a) The game
$$G = \square + \square = 1 - 1 = 0$$
has the left incentive $1 - 0 = 1$ and right incentive $0 - (-1) = 1$.

(b) The game
$$G = \square = \{1 \mid -1\}$$
has the left incentive $1 - \{1 \mid -1\} = 1 + \{1 \mid -1\}$. Combining the disjunctive sum into one game gives that the left incentive is
$$\{0 + \{1 \mid -1\}, 2 \mid 1 - 1\} = \{\{1 \mid -1\}, 2 \mid 0\} = \{2 \mid 0\}.$$
Similarly, the right incentive is also $\{2 \mid 0\}$.

We will see more calculations of incentives early in the next chapter. If the values of game are all numbers, then using incentives is a quick and easy way of deciding which moves are the best.

Problems

1. Find a good description of a *Hasse diagram* online. (Wikipedia will do.)[6]

(a) Draw a Hasse diagram for the integers $\{1, \dots, 10\}$ using the partial order $a \leq b$ when a divides b. (So, $2 \leq 6$ since $\frac{6}{2}$ is an integer, and $2 \leq 2$, but $2 \not\leq 5$.)

(b) Draw a Hasse diagram of the following HACKENBUSH positions using the partial order of games defined in this chapter.

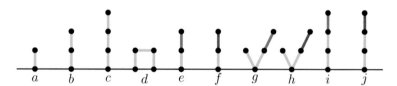

2. In COL

(a) find the canonical form of ●—○—○—○;

(b) find the canonical form of ○—●—○—○;

(c) find the canonical form of ○—○—○—○.

3. In SNORT

(a) find the canonical form of ●—○—○—○;

(b) find the canonical form of ○—●—○—○;

(c) find the canonical form of ○—○—○—○.

4. Let $G = \{\{2 \mid -2\}, \{3 \mid 1\}, \{4 \mid 0\} \mid \{3 \mid -3\}\}$. Show that $G = 1$ by

(a) proving that $G - 1 = 0$;

(b) reducing G to canonical form.

---

[6]This first problem, while tedious, is valuable to gain practice working with the axioms. Consider doing the problem in groups or as an in-class activity.

5. Fix a particular (partial) HACKENBUSH position G, with one edge e unspecified. Four different positions can be obtained depending on whether edge e is blue, green, red, or missing. (If missing, note that other edges might be disconnected from the ground and therefore removed as well.) Give the partial order on these four games, and prove that the partial order is independent of the position. As always, induction is preferred over arguments like, "and then Right continues to ...."

6. The following beautiful DOMINEERING decomposition theorem appears in *ONAG* [Con01] and *WW* [BCG01]:

$$\text{If} \quad \left(G\,\square\right) = \left(G\right) \quad \text{then} \quad \left(G\,\square\,H\right) = \left(G\right) + \left(\square\,H\right)$$

Here, G and H are arbitrary DOMINEERING positions.

For an example application of the theorem, it is not hard to convince yourself that

$$\boxed{} = \boxed{}$$

Then by the theorem

$$\boxed{} = \boxed{} + \boxed{}$$

John Conway's proof in *ONAG* [Con01] is summarized by

$$\left(G\,\square\,H\right) \leq \left(G\right) + \left(\square\,H\right) = \left(G\,\square\right) + \left(\square\,H\right) \leq \left(G\,\square\,H\right)$$

Justify each inequality or equality in the last expression to complete the proof of the DOMINEERING decomposition theorem.

7. (a) Prove that Left's incentive from the TOPPLING DOMINOES position

equals Right's incentive.

(b) State and prove a theorem about how to compare the incentives of two games of the form

$$\uparrow^m \uparrow^n$$

for $m, n \geq 1$. For example, using your theorem it should be possible to compare the incentives from

with those from

Preparation for Chapter 5

The purpose of this (rather fun) exercise is to assure the reader that a number is just a symbol, and context determines its usefulness. Conway's *rational tangles* are a way to describe the various ways in which two strands of rope can be entwined. Place two ends of rope on a table, and label the corners of the table A, B, C, and D in a clockwise fashion. Now, place the ropes on the table, one going from point A to B and one from D to C:

A *twist* is performed by swapping the ends at B and C, passing the end that started at B over the one that started at C. A *rotate* consists of rotating the ends 90 degrees clockwise, $A \to B \to C \to D$. Here are two twists followed by a rotate:

Conway associates with each two-rope tangle a rational number. The starting point is 0. Each twist adds 1, and each rotate takes the negative reciprocal. So, two twists and a rotate yield

$$0 \xrightarrow{T} 1 \xrightarrow{T} 2 \xrightarrow{R} -\frac{1}{2}.$$

He proved that tangles are in one-to-one correspondence with rational numbers. Consequently, you can untangle the above tangle by doing another twist, a rotate, and then two more twists:

$$-\frac{1}{2} \xrightarrow{T} \frac{1}{2} \xrightarrow{R} -2 \xrightarrow{T} -1 \xrightarrow{T} 0.$$

(By the way, rotating 0 yields $-\frac{1}{0} = \frac{1}{0}$.)

Prep Problem 5.1. Determine the sequence of twists and rotates to construct the tangle $\frac{3}{7}$. Next, how do you untangle it with more twists and rotates? Check your answer using two strings or shoelaces.

To the instructor: As alluded to in the previous chapter's suggestions, CUTCAKE and HACKENBUSH from [BCG01, Ch. 2] provide good examples for numbers. Noam Elkies constructs CHESS endgame positions in [Elk96] that include $\uparrow$, $\frac{1}{2}$, $\frac{1}{4}$, ± 1, and $+_1$. If the class has CHESS players, we recommend covering his material as a case study toward the end of this chapter.

From here on, students should be encouraged to use CGSuite to help attack problems (some of which are quite challenging to do entirely by hand) and to translate their observations into clear proofs.

Chapter 5

Values of Games

Y'see it's sort of a game with me. Its whole object is to prove that two plus two equals four. That seems to make sense, but you'd be surprised at the number of people who try to stretch it to five.

Dalton Trumbo in *The Remarkable Andrew*

In this chapter, we begin the process of naming games. On page 76, we assigned names to the four games born by day 1, those being 1, -1, 0, and $*$. Since there are already 1474 distinct games born by day 3, we do not intend to name all games. We will focus on naming the most *important* games, where importance of a game is determined by informal criteria: Does the game appear naturally as a position in played games? Have we proved theorems about the game? Do we know its order relationships with other named games? Do we know how to add the game to other named games?

The reason for assigning specific names to games is not just to provide a shorthand method of referring to them. So, when we choose a name for a game, it should be one that is related to the central properties of that game. Further, the names and notation that we choose should be consistent and easy to remember.

5.1 Numbers

We have already argued that there are good reasons to associate numbers with various DOMINEERING positions (and hence with any other games having the

same structure). Specifically,

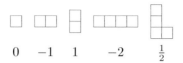

$$0 \quad\quad -1 \quad\quad 1 \quad\quad\quad -2 \quad\quad\quad \tfrac{1}{2}$$

There are other DOMINEERING positions to which we could also easily assign numerical values — for instance, a vertical strip of $2n$ squares represents n moves available to Left and none to Right, so it would seem sensible to declare that it has value n. We will now formalize and continue this development. It turns out that the only numbers that we need to associate with games whose game trees are finite are the *dyadic rationals,* those rational numbers whose denominators are a power of 2. Other numbers do occur as the values of games but require games whose lengths are not bounded in advance. Since we only study short games in this book, all rational numbers that we encounter will be dyadic.

Integers

It is natural to define a game in which Left has n free moves available as having *value* n. Similarly, a game in which Right has n free moves is $-n$. In this way, we define all integers. More formally, for n a positive integer, we define the games

$$0 \overset{\text{def}}{=} \{\,|\,\} \text{ and}$$

$$n \overset{\text{def}}{=} \{n-1 \mid \}.$$

The game $-n$ will be defined as the negative in the sense of the definition given in (4.3) on page 75 of the game n. Remember, in this definition, that n and $n-1$ really stand for "the game whose value is n" and "the game whose value is $n-1$," respectively. So, as a game,

$$2 = \{1 \mid \} = \{\{0 \mid \} \mid \} = \{\{\{ \mid \} \mid \} \mid \}.$$

You can see why it might be preferable just to write 2.

Exercise 5.1. Use the definition of negative on page 79, and confirm that

$$-n = \{\, \mid 1-n \}.$$

Exercise 5.2. Confirm that the DOMINEERING position that is a vertical strip of four empty squares is the game whose value is 2.

Observation 5.3. Let n be an integer and N a game whose value is n.

- If $n = 0$, neither player has a move available in N.

- If $n > 0$, then Left has a move in N to the game whose value is $n - 1$, and Right has no move available in N.

- If $n < 0$, then Right has a move in N to the game whose value is $n + 1$, and Left has no move available in N.

The previous definition already puts us on thin ice — slippery and dangerous. It is slippery in the sense that when we write 10 do, we mean the number 10 or the game whose value is 10? It is dangerous in that we have at present no evidence that the games whose values are integers share any significant properties with the integers. Since we can compare and add games, and we can compare and add integers, we would be in serious trouble if the properties of these games and the corresponding integers differed. As you may well have guessed — they don't. But for peace of mind, we had better confirm that now. We will let you start that process with an exercise:

Exercise 5.4. In this exercise only, write "n" to mean the game whose value is n. Show that "$n + 1$" equals the game sum of "n" and "1". Confirm that the similar property holds for "$n - 1$".

For the remainder of this section, we will try to limit the notational confusion between games and integers by referring to games with capital letters and integers with lowercase ones. Thus, the games A, B, and C will have integer values a, b, and c, respectively. We relegate the proofs to Section 5.9 at the end of this chapter.

Lemma 5.5. *The integer sum $a + b + c \geq 0$ if and only if $A + B + C \geq 0$ as games. (Symmetrically, $a + b + c \leq 0$ if and only if $A + B + C \leq 0$.)*

(Recall that $A + B + C \geq 0$ means that Left wins moving second in $A + B + C$.)

Theorem 5.6. $A + B = C$ *if and only if $a + b = c$.*

In addition to their additive structure, integers are also ordered. As part of justifying the ambiguous notation, we ask the reader to confirm that integer games maintain the same ordering:

Exercise 5.7. Prove that $A \geq B$ if and only if $a \geq b$.

Exercise 5.8. Find the incentives from 0 and from integers $n \neq 0$. (Incentives are defined in Section 4.5 on page 99.) By the way, incentives are never positive or zero.

We now feel comfortable enough with our definition to declare that from now on when we write 17 we might be referring to "the game whose value is 17" (or any other game equal to that specific game) or to the integer 17.

Example 5.9. A HACKENBUSH position that has only blue edges has an integer value equal to the number of edges in the position.

Exercise 5.10. What is the value of the following HACKENBUSH position?

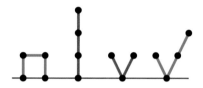

Fractional numbers

Now we will continue identifying games with numbers and, in particular, with the dyadic rational numbers, that is, fractions whose denominators are a power of 2, such as $\frac{1}{2}$, $\frac{3}{8}$, and $\frac{19}{32}$. When the definition of numbers is naturally extended beyond short games, we get the *surreal numbers,* which include the reals, the ordinals, and more [Con01, Knu74].

We can let our previous investigation of DOMINEERING guide the definition:

$$\text{▱} = \left\{ \text{▱} \,\middle|\, \text{▱} \right\} = \{0 \mid 1\}$$

Here, we eliminated a dominated option (Left playing in the top two vertical squares) to produce the canonical form of this position. We argued previously that this value was a pretty good candidate for $\frac{1}{2}$ since

$$0 < \text{▱} < 1 \quad \text{and} \quad \text{▱} + \text{▱} = 1.$$

Exercise 5.11. Try to come up with "reasonable" candidates for $\frac{1}{4}$ and $\frac{1}{8}$ (as abstract games, not as DOMINEERING positions).

We will now give the formal definition of numbers, and then do our ice-dancing trick again.

Definition 5.12. For $j > 0$ and m odd, we define the *number*[1]

$$\frac{m}{2^j} = \left\{ \frac{m-1}{2^j} \,\middle|\, \frac{m+1}{2^j} \right\}.$$

---

[1] A *better* definition of number is any game x such that all $x^L < x < x^R$ and x^L and x^R are numbers. This naturally generalizes to non-short games. That our definition is equivalent for short games will be a consequence of Theorem 5.25.

So, for example, $\frac{19}{32} = \left\{ \frac{18}{32} \mid \frac{20}{32} \right\} = \left\{ \frac{9}{16} \mid \frac{5}{8} \right\}$.

Problem 7 on page 134 asks you to prove that the games given by Definition 5.12 are in canonical form.

Example 5.13. Prove that $\frac{1}{2} + \frac{1}{2} = 1$.

Proof: Recall that in this context $\frac{1}{2}$ stands for $\{0 \mid 1\}$ and 1 stands for $\{0 \mid \}$. It suffices to show that either player wins moving second on the difference game $\frac{1}{2} + \frac{1}{2} - 1$. If either player moves on a $\frac{1}{2}$, the other will respond on the other $\frac{1}{2}$, leaving $1 - 1 = 0$, and so the first player loses. On the other hand, if Right moves -1 to 0, this will leave $\frac{1}{2} + \frac{1}{2}$ from which Left can move to $\frac{1}{2}$, Right can respond to 1, and Left takes the 1 to 0. $\square$

Exercise 5.14. Confirm a few more complicated equalities, such as $\frac{15}{16} + \frac{1}{4} = 1\frac{3}{16}$, until you are comfortable with the definition of numbers. As you do so, assume inductively that simpler values behave as one would expect.

It is important that moves on a number change it to a worse number for the mover. (Left moves make it smaller; Right moves make it larger.) In particular, the *incentive* for either player to move on $\frac{m}{2^j}$ (for m odd and $j > 1$) is $-\frac{1}{2^j}$.

Exercise 5.15. Confirm the last assertion. Note that it is important that $\frac{m}{2^j}$ is in canonical form. Incentives are defined in Section 4.5 on page 99.

We will later strengthen this observation concerning the negative incentives of numbers in two ways. First, the *Number-Avoidance Theorem* (Theorem 6.12 on page 146) states that one should only move on numbers as a last resort. Second, the *Negative-Incentives Theorem* (Theorem 6.15 on page 147) proves that if all the incentives for a game G are negative, then G is a number.

As with integers, in adopting these definitions of numbers, we are suggesting that they add and are ordered as one would expect. Suppose that games A, B, and C have values that are numbers (dyadic rationals) a, b, and c, respectively. As we did for integers, we wish to show that $A + B = C$ *if and only if* $a + b = c$ and that $A > B$ *if and only if* $a > b$. Echoing the proof for integers would be tedious, however, both because of the more complicated definition of numbers and because we want to mix non-integers and integers. The following lemma obtains the results more quickly. Again we adopt the convention of using uppercase letters to stand for "games whose names are numbers" and the corresponding lowercase letters to stand for the numbers themselves.

Lemma 5.16.
$$a + b + c = 0 \iff A + B + C = 0;$$
$$a + b + c > 0 \iff A + B + C > 0;$$
$$a + b + c < 0 \iff A + B + C < 0.$$

Theorem 5.17.

- $A + B = C$ *if and only if* $a + b = c$.

- $A \geq B$ *if and only if* $a \geq b$.

When playing a sum of games, it is generally wisest to avoid playing on numbers. This assertion is codified in two theorems, the *Weak Number-Avoidance Theorem* proved here, and the strong one proved in Chapter 6.

Theorem 5.18. (Weak Number Avoidance) *Suppose that x is a number and G is not. If Left can win moving first on $x + G$, then Left can do so with a move on G.*

Proof: We may assume that x is in canonical form. Rephrased, the theorem claims that if some $x^L + G \geq 0$ then some $x + G^L \geq 0$. Assume that $x^L + G \geq 0$. Since G is not a number (in particular, $G \neq -x^L$), we know that $x^L + G > 0$. So, Left wins moving first on $x^L + G$, and by induction some $x^L + G^L \geq 0$. Since $x > x^L$, we have that $x + G^L \geq 0$. $\qquad\qquad\qquad\square$

The simplest number

The definition of numbers, Definition 5.12, can be generalized to recognize when G is a number even if it is not in canonical form.

Definition 5.19. For numbers $x^L < x^R$, the *simplest number* x between x^L and x^R is defined by the unique number with the smallest birthday strictly between x^L and x^R.

Definition 5.20. (Alternate definition) For $x^L < x^R$, the *simplest number* x between x^L and x^R is given by the following:

- If there are integer(s) n such that $x^L < n < x^R$, then x is the one that is smallest in absolute value.

- Otherwise, x is the number of the form $\frac{i}{2^j}$ between x^L and x^R for which j is minimal. (The reader might recognize this as the longest ruler mark between x^L and x^R.)

Theorem 5.21. *The two definitions of simplest number are well defined and equivalent.*

Before we continue, Problem 17 asks you to prove that if one (or both) of the option sets of a game G is empty then G is an integer. This "integer" will be the simplest number strictly larger than x^L or strictly less than x^R, or 0 if neither exists. In the rest of the section, we can safely assume that both Left and Right have options.

We will first state two lemmas for the reader to prove and then prove the result.

Lemma 5.22. *If $x_1 < x_2$ both have the same birthday, then some number x, with $x_1 < x < x_2$, has a smaller birthday.*

Proof: Problem 8 on page 134 asks you to prove this. $\square$

Lemma 5.23. *The birthdays of all dyadic rational numbers are described as follows:*

- *The birthday of an integer n is the absolute value of n.*

- *The birthdays of $\left(n + \frac{i}{2^j}\right)$ and its negative are $n + j$, where $n \geq 0$ is an integer, i is an odd integer, and $0 < i < 2^j$.*

Exercise 5.24. Prove Lemma 5.23 by induction using the definitions of integer and number.

Proof (of Theorem 5.21): To prove that the definitions are well defined, we need to prove that each produces a *unique* x. For the first definition, if two numbers x_1 and x_2 have the same birthday, then by Lemma 5.22 some x between x_1 and x_2 has a smaller birthday. For the second definition, if n_1 and n_2 are candidate integers with the same (smallest) absolute value, then $n_1 = -n_2$, and so 0 (which is a simplest number between the two) is also a candidate integer. Equivalently, if $\frac{i_1}{2^j}$ and $\frac{i_2}{2^j}$ are candidate numbers, so is $x = \frac{i}{2^j}$ where i is an even number between the two odd numbers i_1 and i_2, and the fraction x reduces to one with a smaller denominator.

To prove that the definitions are equivalent, if an integer lies between x^L and x^R, then the integer with smallest absolute value (by Lemma 5.23) has the smallest birthday. If no integer lies between, then they both lie between two consecutive $n < x^L, x^R < n + 1$, and by Lemma 5.23 both definitions will minimize j. $\square$

Theorem 5.25. *If all options of a game G are numbers and every left option G^L of G is strictly less than every right option G^R of G, then G is also a number. In particular, G is the simplest number x lying strictly between every G^L and every G^R.*

Note that if all the options of G are numbers then, by domination, the canonical form of G will have at most one Left and at most one Right option. However, this theorem and Theorem 5.27, its generalization, do not depend on obtaining the canonical form.

Proof: Fix $G = \{\mathcal{G}^L \mid \mathcal{G}^R\}$ where all $\mathcal{G}^L < \mathcal{G}^R$ are numbers, and let x be the simplest number satisfying $\mathcal{G}^L < x < \mathcal{G}^R$. It suffices to show that the second player wins playing $x - G$. Right playing first can play to some $x - G^L$ or to $x^R - G$. The former is positive since we chose $\mathcal{G}^L < x < \mathcal{G}^R$. For the latter, it cannot be that $x^R < \mathcal{G}^R$, for then $\mathcal{G}^L < x < x^R < \mathcal{G}^R$, but x^R is simpler than x and x was the simplest number in that range. Hence, there is some G^R with $x^R \geq G^R$, and Left's move to $x^R - G^R$ wins.

By a symmetric argument, Right moving second wins, and so $x = G$. □

Exercise 5.26. Determine the values of

(a) $\{\frac{1}{2} \mid 2\}$;

(b) $\{\frac{1}{8} \mid \frac{5}{8}\}$;

(c) $\{-1\frac{27}{64} \mid -1\frac{9}{32}\}$.

Theorem 5.25 can be generalized to handle some cases when the options are not numbers:

Theorem 5.27. *If there is some number x such that $\mathcal{G}^L \lhd x \quad \mathcal{G}^R$, then G is the simplest such x.*

Proof: The proof is nearly identical to that of Theorem 5.25 and is left for the reader as Problem 16. □

Example 5.28. In PUSH, who should win

and how?

First, note that the negative of a PUSH position is obtained by changing the color of each blue and red piece. For Right, moving in

seems bad. Saving a counter that can be pushed or pushing a left counter looks better, but how to decide? Left's best move would appear to be in the last game where she can push two right pieces, but is it a winning move? If the games are all numbers, and at least one contains a fraction, then by the previous exercise the best move will be in the game with the highest denominator.

The position

is clearly worth one move to Left:

$$\boxed{\text{◖}} = \{0 \mid \} = 1.$$

Also,

$$\boxed{\;\;\text{▮}\;} = \{1 \mid \} = 2, \quad \text{and} \quad \boxed{\;\;\;\;\text{▮}} = \{2 \mid \} = 3.$$

With a right piece, we have

$$\boxed{\text{▮}\;\text{▮}} = \left\{\boxed{\text{▮}\;} \;\middle|\; \boxed{\;\;\text{▮}}\right\} = \{1 \mid 2\} = \frac{3}{2}.$$

Also,

$$\boxed{\text{▮}\;\;\;\text{▮}} = \left\{\boxed{\text{▮}\;\text{▮}} \;\middle|\; \boxed{\;\;\;\;\text{▮}}\right\} = \left\{\frac{3}{2} \;\middle|\; 3\right\} = 2,$$

and

$$\boxed{\;\;\text{▮}\;\text{▮}} = \left\{\boxed{\text{▮}\;\text{▮}} \;\middle|\; \boxed{\text{▮}\;\;\;\text{▮}}\right\} = \left\{\frac{3}{2} \;\middle|\; 2\right\} = \frac{7}{4}.$$

Lastly,

$$\boxed{\text{▮}\;\text{▮}\;\text{▮}} = \left\{\boxed{\text{▮}\;\text{▮}} \;\middle|\; \boxed{\;\;\text{▮}\;\text{▮}}, \boxed{\text{▮}\;\;\;\text{▮}}\right\} = \left\{\frac{3}{2} \;\middle|\; \frac{7}{4}, 2\right\} = \left\{\frac{3}{2} \;\middle|\; \frac{7}{4}\right\} = \frac{13}{8}.$$

Therefore, the value of the disjunctive sum of these games is

$$\boxed{\text{▮}\;\text{▮}} + \boxed{\;\;\text{▮}\;\text{▮}} + \boxed{\;\;\text{▮}} + \boxed{\text{▮}\;\text{▮}\;\text{▮}} = -\frac{3}{2} + \frac{7}{4} + (-2) + \frac{13}{8} = -\frac{1}{8},$$

which is a win for Right. Since these values appeared pretty much in canonical form, we can build a table of the positions and their incentives:

Position	Left incentive	Right incentive
$\boxed{\text{▮}\;\text{▮}} = -\frac{3}{2} = \{-2 \mid -1\}$	$-\frac{1}{2}$	$-\frac{1}{2}$
$\boxed{\;\;\text{▮}\;\text{▮}} = \frac{7}{4} = \{\frac{3}{2} \mid 2\}$	$-\frac{1}{4}$	$-\frac{1}{4}$
$\boxed{\;\;\text{▮}} = -2 = \{ \mid -1\}$		-1
$\boxed{\text{▮}\;\text{▮}\;\text{▮}} = \frac{13}{8} = \{\frac{3}{2} \mid \frac{7}{4}\}$	$-\frac{1}{8}$	$-\frac{1}{8}$

Left can only change the sum of the games to a more negative number, but the best of a bad lot is to move in the last summand where the sum only decreases by $\frac{1}{8}$. Right has only one winning move, to move from

$$\boxed{\text{▮}\;\text{▮}\;\text{▮}} \quad \text{to} \quad \boxed{\;\;\text{▮}\;\text{▮}}$$

which changes the sum to 0 — every other move leaves a positive game that Left will win.

When a number appears with a reversible option (and is not in canonical form), then the incentive could be different. For example, the left and right incentives from

$$\boxed{\text{▮}\;\;\;\text{▮}} = \left\{\frac{3}{2} \;\middle|\; 3\right\} = 2$$

are $-\frac{1}{2}$ and -1, respectively. However, if all you seek is *some* component with a winning move (if it exists), you may assume that each component is in canonical form.

While we do not yet have a formula to easily compute the value of any PUSH position, a close variant called SHOVE has been solved.

5.2 Case Study: Shove

The game of SHOVE is played on one or more strips of squares. Each square can be empty or it might contain a blue or red counter. Left moves by selecting a blue counter and moving it, along with all the counters to its left on the same strip, left one square. Counters can fall off the left-hand end of a strip. Right moves by selecting a red counter and moving it, along with all the counters to its left, left one square. Note that counters always move leftward.

Exercise 5.29. Play SHOVE[2] with a classmate or friend from the following start position (you might want to use coins as counters):

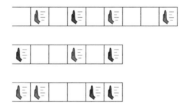

In order to analyze this game, we need only consider positions consisting of a single strip, as the multiple-strip positions are sums of such games. Though not at all obvious from the rules, there is a simple recipe for calculating the value of any SHOVE position.

First, number the squares $\{1, 2, \ldots\}$ from the left. Assume that the right-most piece is blue and is on square n. If the piece 2nd from the right is also blue, or if there are no red pieces, then the value of G is n plus the value of the position achieved by removing the piece on square n. For example,

The justification for this is that, faced with a choice between moving either her rightmost piece or the piece second from the right, Left always prefers to move the piece second from the right. You can verify this by a direct comparison of the games resulting from each of the possible two moves, or more elegantly observing that the latter move "preserves future options" for Left. This general

---

[2]In this text, SHOVE pieces look almost like PUSH pieces, only with more lines behind the hand.

technique (which amounts to eliminating from consideration a certain type of dominated option) might well be called the *Don't Burn Your Bridges Principle*.

After repeated application of this rule, we will either obtain an empty position (and know that the value of the position was an integer) or reach a position in which the two rightmost counters are of opposite colors.

In a position in which the rightmost piece is blue and the second from the right is red, a piece at position n with c counters to its right contributes $\frac{n}{2^c}$ multiplied by the appropriate sign.

Each counter below is labeled with its value as given by the recipe:

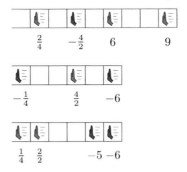

$$\frac{2}{4} \qquad -\frac{4}{2} \qquad 6 \qquad\qquad 9$$

$$-\frac{1}{4} \qquad\qquad \frac{4}{2} \qquad -6$$

$$\frac{1}{4} \quad \frac{2}{2} \qquad\qquad -5 \ -6$$

More formally, any n-piece SHOVE position may be defined by two functions p and c from $\{1, 2, \ldots, n\}$ to $\mathbb{N}$, where

$$p(i) = \text{position of } i^{\text{th}} \text{ piece from the left;}$$

$$c(i) = \begin{cases} 1 & \text{if the } i^{\text{th}} \text{ piece is blue,} \\ -1 & \text{if the } i^{\text{th}} \text{ piece is red.} \end{cases}$$

Additionally, define $r(i)$ to be the number of pieces strictly to the right of the ith piece up to and including both pieces of the last color alternation. In particular, $r(i) = 0$ only if the ith piece and all pieces to its right are the same color.

Theorem 5.30. *Adopting the preceding notation, the value of a* SHOVE *position is*

$$x = \sum_{1 \leq i \leq n} c(i) \frac{p(i)}{2^{r(i)}}.$$

Proof: One of Left's options (as we will see, her best option) is to move her leftmost piece j at position $p(j)$. By induction, that move is to a position whose value is

$$\sum_{1 \leq i \leq j} c(i) \frac{p(i) - 1}{2^{r(i)}} + \sum_{j < i \leq n} c(i) \frac{p(i)}{2^{r(i)}} = x + \sum_{1 \leq i \leq j} c(i) \frac{-1}{2^{r(i)}}$$

$$= x - \frac{1}{2^{r(1)}}.$$

Note that this expression is correct even in the case where $p(1) = 0$. For the last equality, since j is the leftmost blue piece, $c(j) = 1$, while $c(i) = -1$ for $i < j$. Left's options to move any other piece, say $j' > j$, is inferior, for then the last summation evaluates to a quantity $< x - \frac{1}{2^{r(1)}}$ since $c(j') = c(j) = 1$.

By a symmetric argument, Right's best option is to $x + \frac{1}{2^{r(1)}}$. Hence, the value of the shove position is given by

$$\left\{ x - \frac{1}{2^{r(1)}} \;\middle|\; x + \frac{1}{2^{r(1)}} \right\} = x$$

because x is the simplest number between the two.   □

You might well wonder, "Where on earth did that formula come from?" The answer is fairly simple. When we begin to analyze a new game, we always start by looking at simple positions. In SHOVE the simplest positions are those in which there is only one counter — their values are rather trivial. Next, we might well observe that positions in which all the counters are the same color have equally trivial values. So, the next cases to consider are positions with two counters, one of each color. Starting with the case of a red counter on square 1 and a blue counter somewhere to its right, we find red counter to be worth half a move to Right. When the red counter is on square 2, we find that it was worth a full move. And so on. At this point we might well introduce CGSuite to consider slightly more complex positions. We soon observe that the values of SHOVE positions always seem to be numbers, and we make the observation that the pieces to the right of the final color alternation could be removed (with suitable compensation). Then, we might look at the effect on the value of removing the leftmost (or rightmost) counter — and eventually come up with the formula at the end of the proof, along with a fair degree of confidence in its correctness. After that, it's just a matter of slogging through the algebra to provide the proof that the formula is correct.

Exercise 5.31. For the following SHOVE position from the start of this section, find the incentive for each move and then identify all the best first moves for each player:

5.3 Stops

Suppose that two players play a game and agree to stop playing as soon as the value of the position that has been reached is a number. Left attempts to have this stopping position be as large as possible, and Right wants it to be as small as possible. The number arrived at when Left moves first is called the *left stop* of the original game, while that reached when Right moves first is called the *right stop*. The following definition puts this more formally.

Definition 5.32. The *left stop* and *right stop* of a game G are denoted by $\mathbf{LS}(G)$ and $\mathbf{RS}(G)$, respectively. They are defined in a mutually recursive fashion:

$$\mathbf{LS}(G) = \begin{cases} G & \text{if } G \text{ is a number,} \\ \max(\mathbf{RS}(G^L)) & \text{if } G \text{ is not a number;} \end{cases} \tag{5.1}$$

$$\mathbf{RS}(G) = \begin{cases} G & \text{if } G \text{ is a number,} \\ \min(\mathbf{LS}(G^R)) & \text{if } G \text{ is not a number.} \end{cases} \tag{5.2}$$

Exercise 5.33. Compute the left and right stops of each position marked with a box below. Rest assured that the game is in canonical form, so none of the interior nodes are numbers in disguise:

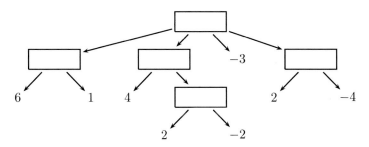

Theorem 5.34. *For any game* G, $\mathbf{LS}(G) \geq \mathbf{RS}(G)$.

Proof: If G is a number, then $\mathbf{LS}(G) = \mathbf{RS}(G)$. If G is not a number, suppose that the theorem is false and that $\mathbf{LS}(G) < \mathbf{RS}(G)$. Then, there is some number x where $\mathbf{LS}(G) < x < \mathbf{RS}(G)$. To complete the proof, we will show that $G = x$, contradicting our assumption that G is not a number. To show that, we will show that $G - x = 0$, i.e., the first player has no winning move on $G - x$. When playing $G - x$, the Weak Number-Avoidance Theorem asserts that if there is a winning move, it is on G. So, since we are assuming that G is not a number, without loss of generality, both players play on G until it reaches a number. When Left moves first, the maximum she can achieve playing this way is $\mathbf{LS}(G)$, and when Right moves first, the minimum he can achieve is $\mathbf{RS}(G)$. Neither are good enough for the first player to achieve a win on $G - x$. □

Stops give us a (sometimes) handy way of comparing two games.

Lemma 5.35. *Let* x *be a number and* G *and* H *be games. If* $\mathbf{RS}(G) > x$, *then* $G > x$. *If* $\mathbf{RS}(G) > \mathbf{LS}(H)$, *then* $G > H$.

Proof: Consider playing $G - x$. By the Number-Avoidance Theorem, neither player plays in x until G is reduced to a number. Under alternating play, the best result that Right can achieve moving first is $\mathbf{RS}(G) - x > 0$, so he loses moving first. Left moving first can achieve $\mathbf{LS}(G) - x \geq \mathbf{RS}(G) - x > 0$. Therefore, $G - x > 0$

If $\mathbf{RS}(G) > \mathbf{LS}(H)$, then there is some dyadic rational y where $\mathbf{RS}(G) > y > \mathbf{LS}(H)$. By the first part, $G > y$ and $y > H$, therefore $G > H$. □

Exercise 5.36. Show that if $G = \{7 \mid \{1 \mid -1\}\}$ and $H = \{\{4 \mid -2\} \mid -3\}$, then $G > H$. (You may assume that they are in canonical form.)

Exercise 5.37. Let $G = \{\{4 \mid -2\}, \{3 \mid 1\} \mid \{3 \mid -1\}, \{2 \mid 0\}\}$. Show that G is not in canonical form.

One can adopt other definitions of when to stop a game. For example, one could define the *integer stops* of a game by replacing the word "number" with "integer" in Definition 5.32. The analogy to Theorem 5.34 is the following:

Corollary 5.38. *The difference between the integer left stop and the integer right stop of any game G is at least -1. Further, when the difference is -1, the person who moves first on G also moves last on G to reach the stop.*[3]

Proof: Suppose that Left is playing first. By playing toward $\mathbf{LS}(G)$, she can ensure an integer stop of at least $\lfloor \mathbf{LS}(G) \rfloor$, while Right can ensure an integer stop of at most $\lceil \mathbf{LS}(G) \rceil$. The former can occur only if the fractional part of $\mathbf{LS}(G)$ is at most $\frac{1}{2}$ and it is Left's turn to play when the stop is reached. Similar arguments apply to the Right integer stop. However, we also know that $\mathbf{LS}(G) \geq \mathbf{RS}(G)$. So, the only way that the difference between the integer stops could be negative would be for Left to be forced to an integer stop of $\lfloor \mathbf{LS}(G) \rfloor$, while Right can attain one of $\lceil \mathbf{RS}(G) \rceil$ (and furthermore that $\mathbf{LS}(G)$ and $\mathbf{RS}(G)$ lie between the same pair of integers). Observe that the difference in integer stops can be -1 only if $\mathbf{LS}(G)$ and $\mathbf{RS}(G)$ are numbers strictly between the same consecutive integers, n and $n + 1$ and Left is forced to move first on $\mathbf{LS}(G)$, while Right moves first on $\mathbf{RS}(G)$. □

5.4   A Few All-Smalls: Up, Down, and Stars

We have already named one game that is not a number, that being $* = \{0 \mid 0\}$. We saw this game as one of the four games born on day 1. There are many important games that are nearly 0 but are not numbers.

---

[3]The last sentence of the corollary will be relevant when we prove Theorem 9.32 several chapters hence.

Definition 5.39. A game G is *infinitesimal* if $-x < G < x$ for all positive numbers x.

Example 5.40. We will show that $*$ is infinitesimal. Since $*$ is incomparable with 0 (the first player wins from $*$), $*$ cannot be a number. Note that $* = -*$, so by symmetry it suffices to show that $* < x$ for positive numbers x (i.e., Left wins moving first or second on $x - * = x + *$). From $x + *$, Left wins moving first by moving to x. On the other hand, Right's choices for his first move from $x + *$ are not terribly palatable. By the Weak Number-Avoidance Theorem (Theorem 5.18), if Right had a winning first move, it would have to be in $*$. However, his only available move there leaves $x > 0$, which Left will now win moving first. Since Right has no winning first move, but Left does, $x + * > 0$.

There are many stars that arise out of NIM that we'll explore further in Chapter 7.

We now proceed to define two more infinitesimals:

Definition 5.41. The games *up* and its negative, *down*, are given by

$$\uparrow \ \overset{\text{def}}{=}\ \{0 \mid *\};$$
$$\downarrow \ \overset{\text{def}}{=}\ \{* \mid 0\}.$$

We write $\Uparrow = \uparrow + \uparrow$ for "double-up," $\Uparrow\uparrow = \uparrow + \uparrow + \uparrow$ (triple-up), $\Downarrow = \downarrow + \downarrow$ (double-down), etc.

For example, in CLOBBER, ▣▣▣ $= \uparrow$ and ▣▣▣ $= \downarrow$.

Observation 5.42. The game $\uparrow$ is a positive infinitesimal. Correspondingly, $\downarrow$ is a negative infinitesimal.

Proof: We wish to show that $0 < \uparrow < x$ for numbers $x > 0$. Left wins moving first or second on $\uparrow$, so $\uparrow > 0$. Assume that $x > 0$ is a number. In the game $x - \uparrow$, again by Theorem 5.18 if Right had a winning first move, it would have to be in $-\uparrow$. But his only move there is to 0, leaving x, and so Left wins. Left can win moving first from $x - \uparrow$ to $x - *$ (which is positive since $*$ is infinitesimal). So, $x > \uparrow$. □

Definition 5.43. G is *all-small* if every position H in G has the property that Left can move from H if and only if Right can.

The games $*$ and $\uparrow$ are both all-small games, while the only all-small number is 0.

Observation 5.44. A game G is all-small if and only if either

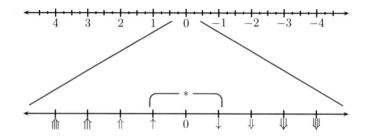

Figure 5.1. Multiples of ↑ are all infinitesimal (i.e., they lie between $+x$ and $-x$ for any positive number x). The game $*$ is incomparable with ↑, 0, and ↓ but lies strictly between ⇑ and ⇓.

1. $G = 0$, or

2. $\mathcal{G}^L$ and $\mathcal{G}^R$ are non-empty and every element of each is all-small.

Theorem 5.45. *Every all-small game is infinitesimal.*

Proof: Problem 14 asks you to prove this.   □

Sums made from ↑ and $*$

We now know how the games ↑ and $*$ compare with all numbers. How do they compare with one another? How about sums of these games?

Observation 5.46.
$$\downarrow < 0 < \uparrow$$
$$\downarrow \parallel * \quad \uparrow$$
$$\Downarrow < * < \Uparrow$$

We leave the proof as a short exercise:

Exercise 5.47. Prove Observation 5.46. We have already shown the first inequality in Observation 5.42. Confirm the other two by playing the games $\uparrow - *$ (which equals $\uparrow + *$) and $\uparrow + \uparrow - *$.

Figure 5.1 summarizes the relative ordering of numbers, ups, and star. In combinatorial game theory, we draw the number line backward, with the positive numbers to the left so as to reinforce the convention that Left is positive.

On notation: When addition is implicit

In grade school, some of us learned[4] that $1\frac{1}{2}$ is shorthand for $1 + \frac{1}{2}$. We will adopt the same convention, that when concatenating named games, we mean

---

[4]Since the 1st edition, Adam Atkinson reports that not all countries use this shorthand notation.

to add them. We always list numbers, then ↑s, and then ∗. So,

$$2\!\Uparrow\!* = 2 + \Uparrow + *,$$

but we would *not* write ∗↑. We will later define a game called ∗2, and it will not equal 2∗ = 2 + ∗.

Canonical forms of $n\!\cdot\!\uparrow$ and $n\!\cdot\!\uparrow\!*$

We will first compute the canonical forms of ↑∗ and ⇑. Since ↑∗ is an option of ⇑, it is best to work on ↑∗ first. Here is the game tree for ↑∗:

Right's move to ↑ is dominated by 0, and there are no other dominated options. As we check for reversible options, note that ↑∗ is incomparable with 0, but that ∗ < ↑∗ since ↑ > 0. So, the move from ↑∗ to ↑ reverses through ∗ to 0:

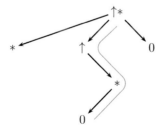

There are no further dominated or reversible options, and we arrive at the canonical form of ↑∗ = {0, ∗ | 0}:

We next compute the canonical form of ⇑. Either player can move on one of the ↑s: in Left's case moving the ↑ to 0, and in Right's case moving ↑ to ∗. So, ⇑ = {↑ | ↑∗}. Clearly, there are no dominated options, for each player has only one option. Right's move to ↑∗ = {0, ∗ | 0} does not reverse, for ⇑ > 0 and ⇑ > ∗. However, Left's move to ↑ reverses through ∗ to 0, for ⇑ > ∗:

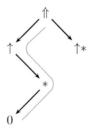

We arrive at the canonical form of ⇑:

In a similar fashion, we can mechanically compute the canonical forms of ⇑∗, ⇑⇑, ∗, and so forth. A pattern quickly emerges:

$$
\begin{aligned}
\uparrow &= \{0 \mid *\}, & \uparrow* &= \{0, * \mid 0\}, \\
\Uparrow &= \{0 \mid \uparrow*\}, & \Uparrow* &= \{0 \mid \uparrow\}, \\
&= \{0 \mid \Uparrow*\}, & * &= \{0 \mid \Uparrow\}, \\
&= \{0 \mid \quad *\}, & * &= \{0 \mid \quad\}.
\end{aligned}
$$

We denote

$$
n \cdot g = \begin{cases}
0 & \text{if } n = 0, \\
\overbrace{g + g + \cdots + g}^{n \text{ times}} & \text{if } n > 0, \\
(-n) \cdot (-g) & \text{if } n < 0.
\end{cases}
$$

So, for example, $3 \cdot \uparrow = \quad$ and $-3 \cdot \uparrow = \Downarrow$.

Theorem 5.48. *For $n \geq 1$, the canonical forms of $n \cdot \uparrow$ and $n \cdot \uparrow *$ (parsed as $(n \cdot \uparrow) *$) are given by*

$$
n \cdot \uparrow = \{0 \mid (n-1) \cdot \uparrow *\};
$$

$$
n \cdot \uparrow * = \begin{cases}
\{0 \mid (n-1) \cdot \uparrow\} & \text{if } n > 1, \\
\{0, * \mid 0\} & \text{if } n = 1.
\end{cases}
$$

Symmetrically,

$$
n \cdot \downarrow = \{(n-1) \cdot \downarrow * \mid 0\},
$$

$$
n \cdot \downarrow * = \begin{cases}
\{(n-1) \cdot \downarrow \mid 0\} & \text{if } n > 1, \\
\{0 \mid 0, *\} & \text{if } n = 1.
\end{cases}
$$

Proof: Problem 18 asks you to prove this.   □

It might be instructive to compare the game trees for 2^{-n} and $n \cdot \uparrow$. For example,

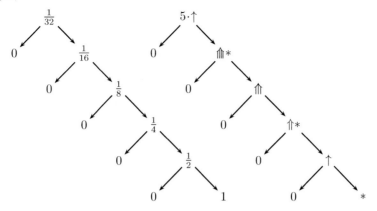

Values with similar game trees can differ dramatically in their algebraic behavior. In particular, note that though the basic structure of these two trees is the same, in the first tree as we proceed along the long branch to the right the values increase, while in the second tree they decrease.

Ups in Clobber

Let us investigate some of the simplest CLOBBER positions, those with a single red stone and a row of blue stones: ▨▨▨▨▨⋯▨. We can build a short list of positions and a pattern quickly emerges:

$$▨ = 0$$
$$▨▨ = \{0 \mid 0\} = *$$
$$▨▨▨ = \{0 \mid *\} = \uparrow$$
$$▨▨▨▨ = \{0 \mid \uparrow\} = \Uparrow *$$

Pictorially,

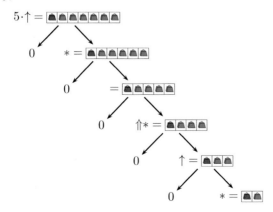

Exercise 5.49. Formalize the pattern by giving an expression for the CLOB-BER position ▲▲▲▲▲ ⋯ ▲ with n blue stones. Inductively prove that your expression holds true for all n.

5.5 Switches

The positions $\{y \mid z\}$ for y and z numbers with $y > z$ are called *switch games* or *switches*. We can normalize any such game by rewriting

$$\{y \mid z\} = a + \{x \mid -x\} = a \pm x,$$

where $a = (y + z)/2$, $x = (y - z)/2$, and $\pm G$ is shorthand for $\{G \mid -G\}$.

Exercise 5.50. Prove that $\{y \mid z\} = a + \{x \mid -x\} = a \pm x$ as advertised above. (We will later see that this is a special case of the *Number-Translation Theorem* on page 146.)

In the switch $\pm x$, the person who plays gets x points. In contrast with numbers, players are eager[5] to play on switches, making them *hot* games.[6]

Naturally, $a \pm x \pm y \pm z$ is shorthand for $a + \pm x + \pm y + \pm z$.

Exercise 5.51. Evaluate $\pm x \pm x$ where $x > 0$ is a number.

Exercise 5.52. Consider the game

$$G = a \pm x \pm y \pm z,$$

where $x > y > z > 0$ and a are all numbers.

(a) Play the game (perhaps with a classmate) with various values of a, x, y, and z.

(b) For what values of a does Left win moving first? Express your answer as an inequality in x, y, and z. How about moving second?

(c) Generalize the last answer to

$$G = a \pm x_1 \pm x_2 \pm x_3 \cdots \pm x_n.$$

While it is easy to evaluate the winner in sums of switches, their canonical forms grow large. For instance,

$$5 \pm 3 \pm 2 \pm 1 = \left\{ 11 \mid 9 \,\middle\|\, 7 \mid 5 \,\middle\|\middle\|\, 5 \mid 3 \,\middle\|\, 1 \mid -1 \right\}.$$

Exercise 5.53. The TOPPLING DOMINOES position with m blue dominoes has value m; if there are m blue dominoes followed by n red dominoes, then the value is $\{m - 1 \mid -(n - 1)\}$. What is the value of the game if there are m blue dominoes followed by n red dominoes and then p blue where $m \geq p$?

---

[5] Well, pretty eager. A game can never have positive incentive.

[6] Hot games are defined in Definition 6.17 and explored in detail in Chapter 8.

Hard switch-like games

One can generalize $\pm x$ to obtain games of the form $\{x \parallel y \mid z\}$ where $x \geq y \geq z$ are numbers. Best play on sums and differences of games of this form is very hard, indeed. Computer science has techniques for formally proving that a problem is *computationally hard,* and the following theorem is for those readers who can appreciate such results.

Theorem 5.54. *Determining whether Left can win playing first on sums and differences of games of the form $\{x \parallel y \mid z\}$ is PSPACE-complete.*

If you know about PSPACE-completeness, read on a bit longer. The reader who does not can safely skip to the next section.

Theorem 5.54 can be used to prove that other games are PSPACE-hard, but some care is required. In general, one wishes to show that $\{x \parallel y \mid z\}$ appears in the game. Unfortunately, most of the time, if the game does appear, it requires a board size that is polynomial in x (not merely polynomial in the number of bits in x) and hence requires an exponential time reduction. It is still an important open question to prove Theorem 5.54 when x, y, and z are integers specified in unary.

Not all is lost, however. Let β exceed the number of bits in x, y, and z. Although the game tree for integer x has a number of positions that are exponential in β, one can move the decimal point, for the game $\frac{1}{2^{\beta}}$ has only a polynomial number of positions in β. If you can then exploit the fact that $\frac{x}{2^{\beta}}$ can be decomposed into a sum of positions of the form $\frac{1}{2^{\beta'}}$, you are done.

In summary, if you can construct $\left\{\frac{x}{2^{\beta}} \parallel \frac{y}{2^{\beta}} \mid \frac{z}{2^{\beta}}\right\}$ on a game board polynomial in size in β, you can prove that the game is hard.

For a proof of Theorem 5.54, and an application to proving that GO endgames are PSPACE-hard, see [Wol02].

5.6   Case Study: Elephants & Rhinos

Who wins in the following game of ELEPHANTS & RHINOS?

The first observation to make is that the board breaks up into smaller boards: reading from left to right, if there is a ● followed by ● then pieces to the left of and including this ● never interact with the pieces to its right. The given game breaks up into the disjunctive sum of four boards:

If any summands consists only of ●s (or of ●s), then computing their values is simply a matter of counting moves. The only positions remaining that require analysis are those in which ●s are all to the left of ●s. Some are easy: On the first board, ● has 6 moves and ● has 1, and the value is $6 - 1 = 5$. On the third board, both players have a move to 0, so its value is $*$.

Fix a position G in which the two *central pieces*, one elephant and one rhino, are separated by s spaces. To the left of these central pieces lie $n_\bullet$ additional elephants, and to the right are $n_\circ$ rhinos. (The total number of pieces is $n_\bullet + n_\circ + 2$.) Let $m_\bullet$ be the maximum number of moves that the elephants can make without moving the center elephant and likewise $m_\circ$ for the rhinos.

Moving the center elephant gains $n_\bullet$ moves (each of the other elephants can move one space extra), while also restricting the rhinos. So, it seems clear that moving the center piece dominates all other moves. In fact, the value of this game can be found by assuming that the central pieces take turns approaching one another until they come to a standstill. We will proceed to formalize this.

Define

$$x = m_\bullet - m_\circ + \left\lfloor \frac{s}{2} \right\rfloor (n_\bullet - n_\circ).$$

Exercise 5.55. Suppose that the two players each make $\lfloor \frac{s}{2} \rfloor$ moves with their central beasts so that they are separated by 0 or 1 square. Convince yourself that x represents the remaining difference between Left's and Right's available moves by non-central beasts.

Claim 5.56. *Using the above notation,*

$$G = \begin{cases} x & \text{if } s \text{ is even,} \\ \{x + n_\bullet \mid x - n_\circ\} & \text{if } s \text{ is odd.} \end{cases}$$

Proof: Suppose that $s = 0$; the result is trivial. Using induction, if $s > 0$ is even, Left can either move the central ● to $\{x + n_\bullet + n_\circ \mid x\}$ or a non-central ● to $x - 1$. The former move dominates, and applying similar considerations to Right, we get

$$G = \left\{ (x + n_\bullet + n_\circ) \mid x \,\middle\|\, x \mid (x - n_\bullet - n_\circ) \right\} = x.$$

If s is odd, Left can move a central ● to $x + n_\bullet$ or (possibly) a non-central ● to $\{x + n_\bullet - 1 \mid x - n_\circ - 1\}$. The former moves dominates, and similarly Right's option to $x - n_\bullet$ dominates his other options. □

In the second component of our motivating example, ●▭▭●▭▭●▭●▭● , we have $s = 1$, $m_\bullet = 2$, $m_\circ = 5$, $n_\bullet = 1$, and $n_\circ = 2$, so $x = 2 - 5 + \lfloor \frac{1}{2} \rfloor(-1) = -3$ and the value is $\{-2 \mid -5\}$. In the fourth component ●▭▭●▭▭●▭▭▭●▭●●● , $s = 5$, $m_\bullet = m_\circ = 6$, $n_\bullet = 2$, and $n_\circ = 3$, so $x = 6 - 6 + \lfloor \frac{5}{2} \rfloor(2 - 3) = -2$ and the game has value $\{0 \mid -5\}$. Therefore,

$$= 5 + \{-2 \mid -5\} + * + \{0 \mid -5\}$$
$$= \left\{3 \mid 0 \,\middle\|\, -2 \mid -4\right\} + *.$$

Right's only losing initial moves are to move a non-central ●. Left, however, can only win by moving the central ● in the last component.

If you do not wish to calculate the value of each position, an optimal strategy is to select the component, among all the components in which the central beasts can move, that contains the largest number of beasts, and to move your central beast in that component. If there are no central beasts, move any legal piece.

Exercise 5.57. Convince yourself of the last assertion. In particular, what is the incentive for each central and non-central move?

5.7 Tiny and Miny

There are two more infinitesimals of note, *tiny-G* and its negative *miny-G*, which are denoted $+_G$ and $-_G$, respectively:

$$+_G = \left\{0 \,\middle\|\, 0 \mid -G\right\},$$

$$-_G = \left\{G \mid 0 \,\middle\|\, 0\right\}.$$

Usually, G is a positive number, but the properties of these infinitesimals remain unchanged whenever G exceeds some positive number.

Exercise 5.58. Let G be a game that exceeds all negative numbers. Confirm that $+_G$ is a positive infinitesimal. (What happens when G is equal to a negative number?)

A game $g > 0$ is *infinitesimal with respect to* $h > 0$ if, for all integers n, $n \cdot g < h$. So, for example, a positive game is infinitesimal if and only if it is infinitesimal with respect to 1.

Theorem 5.59. $+_x$ *is infinitesimal with respect to* ↑ *for any positive number* x.

Proof: It suffices to show that $\uparrow + n \cdot +_x \geq 0$ for all n. Whatever Right's first move, Left "attacks" any remaining $+_x$ moving it to $\{x \mid 0\}$. If Right responds locally, i.e., by choosing the option 0 from $\{x \mid 0\}$, then Left continues in the same fashion (or wins because only ↑ remains). If Right fails to respond locally,

Left's move to x dominates all other positions. Or, Left can play arbitrarily on the other positions, ignoring x until it is the only game remaining. If, in the meantime, Right has played on x, that play only increases x. Once all the $\rightharpoonup_x$s are gone, it is Left's move on either $\uparrow$ or $*$ (if Right's first move was on $\uparrow$), and Left moves to 0. □

Exercise 5.60. Confirm that if $x > y > 0$ are numbers, then $\boldsymbol{+}_x$ is infinitesimal with respect to $\ _y$.

Tinies and minies are the purest examples of threats: If Right is allowed two moves in a row from $\ _x$, he can cash in on x "points." Despite Right's saved-up threat, Left still wins moving first or second, but the threat makes Left's advantage minuscule. In the game of GO (which is not a pure combinatorial game due to the possibility of loops), this threat becomes more significant because these threats can act as *ko threats*, which can be significant for Red.

To emphasize the relationships between important infinitesimals so far, Figure 5.2 on page 129 expands on Figure 5.1.

5.8   Case Study: Toppling Dominoes

While playing sums of TOPPLING DOMINOES positions can be quite challenging, most of our favorite named values appear:

$$\text{⫿⫿⫿⫿} = 8,$$

$$\text{⫿⫿⫿⫿} = \pm 3,$$

$$\text{⫿⫿} = *,$$

$$\text{⫿⫿⫿} = \uparrow,$$

$$\text{⫿⫿⫿⫿} = \ _4,$$

$$\text{⫿⫿⫿⫿} = \frac{1}{16},$$

$$\text{⫿⫿⫿⫿} = *4 \quad \text{(Chapter 7 introduces } *n\text{)}.$$

In particular:

- Integer n is n consecutive blue dominoes $(n \geq 0)$.

- Switch $\{m \mid -n\}$ is $m + 1$ blue dominoes followed by $n + 1$ red dominoes $(m, n \geq 0)$.

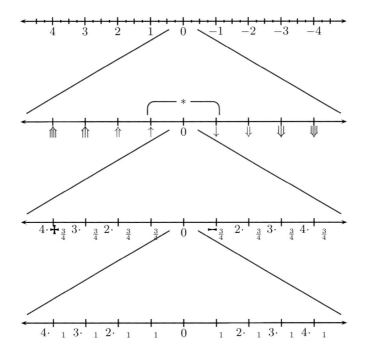

Figure 5.2. While ↑ is infinitesimal, $\Uparrow_x$ is infinitesimal with respect to ↑, and $\Uparrow_y$ is infinitesimal with respect to $\Uparrow_x$, for numbers $y > x > 0$.

- While

  is ↑, we do not know how to create ↑∗ or ⇑ other than as a sum. For instance,

$$\text{🁒🁒🁒🁒} \quad \text{🁒🁒🁒🁒} \quad \text{🁒🁒} = \uparrow + \uparrow + * = \Uparrow *.$$

- $\Uparrow_n$ is $n + 2$ red dominoes sandwiched between 2 blue ones $(n \geq 0)$.

- $\frac{1}{2^n}$ is an alternating sequence of $n + 1$ blue dominoes with n red ones.

- $*n$ is an alternating sequence of n blue and red dominoes.

All of the above are easily confirmed by induction. While Left can move from $\frac{1}{2^n}$ to games in the last case, you need not know that the latter games have value $*n$; you need only know that they are infinitesimal and incomparable with 0 to prove that $\frac{1}{2^n}$ is the correct value. Right's move to $\frac{1}{2^{n-1}}$ is the dominant one (toppling a red and blue domino), while Left's moves that fail to topple all the dominoes reverse out. Here are a few observations and results to help you make sense of the situations.

In what follows we will denote TOPPLING DOMINOES positions by strings of Ls and Rs representing blue and red dominoes, respectively. Greek letters denote arbitrary strings of this type (including the empty one). In this form, for a position $\alpha L\beta$, the two moves available to Left that correspond to toppling the domino designated L leave remaining positions α (toppling to the east) or β (toppling to the west). Determining the outcome class of any string is easy.

Lemma 5.61. *A* TOPPLING DOMINOES *position's outcome class is completely determined by its end dominoes:*

- *The empty string has value* 0 *(and is the unique position of value* 0*);*

- $L > 0$, *and* $L\alpha L > 0$ *for any string* α;

- $R < 0$, *and* $R\alpha R < 0$ *for any string* α;

- $L\alpha R \not\geq 0$ *and* $R\alpha L \not\geq 0$ *for any string* α.

Proof: From $L\alpha L$, any opening move for Right annihilates at most one of the two endpoints, and hence Left has at least one winning response to 0. By symmetry, any string both of whose ends are Rs has negative value. Finally, from $L\alpha R$ (or $R\alpha L$) both players have moves to 0, so such positions belong to $\mathcal{N}$.   □

The next results shed light that lead toward winning strategies in more complex positions.

Lemma 5.62. *For any strings* α *and* β, *every Left move from* $L\alpha L$ *to some* $L\beta R$ *is reversible through* 0.

Proof: Since $L\alpha L > 0$, any Left move to $L\beta R$ reverses through Right's response to 0, which topples the remaining dominoes.   □

The next result shows that in any difference $G - H = 0$, if Left moves to any $G^L - H$, then Right's winning response cannot be to topple in the same direction in G.

Lemma 5.63. (Sandwich Lemma) *Let* G *be a* TOPPLING DOMINOES *position, and suppose that* G^L *is obtained from* G *by toppling east (respectively west). Suppose that* G^{LR} *is obtained from* G^L *by toppling in the same direction. Then,* $G^{LR} \, \rhd \, G$.

Proof: Assume (by symmetry) that both moves toppled east. Write $\alpha = G^{LR}$ so that $G^L = \alpha R \beta$ for some β and $G = \alpha R \beta L \gamma$ for some γ. Note that $\alpha = G^{LR}$ is available directly as a Right option of G, so $\alpha \rhd G$.   □

Intuitively, the best move in a block of dominoes that are all of the same color is to topple one of the end ones away from the block. Here's how to prove it.

Lemma 5.64. *For any* α, $L\alpha > \alpha$.

Proof: In $L\alpha - \alpha$ Left can win by moving to $\alpha - \alpha = 0$. When Right moves first, Left can follow the One-Hand-Tied Principle combined with Tweedledum-Tweedledee by ignoring the designated L and playing as if in $\alpha - \alpha$.   □

It is easy to now obtain the domination result.

Corollary 5.65. *Within any block of adjacent blue dominoes, the two moves that topple an end domino away from the block dominate all other moves within the block.*

Exercise 5.66.   With $a \geq b \geq 0$ and $c \geq 0$, construct a TOPPLING DOMINOES position with value $\{a \parallel b \mid -c\}$.[7]

Exercise 5.67. Construct $\frac{3}{8}$ in TOPPLING DOMINOES.

5.9   Proofs of Equivalence of Games and Numbers

In this section we prove theorems from Section 5.1. The proofs are not hard, but they would have been a bit distracting to the development.

Lemma 5.68. (Lemma 5.5 restated with proof) *The integer sum* $a+b+c \geq 0$ *if and only if* $A + B + C \geq 0$ *as games. (Symmetrically,* $a + b + c \leq 0$ *if and only if* $A + B + C \leq 0$.)

(Recall that $A+B+C \geq 0$ means that Left wins moving second in $A+B+C$.)

Proof: First, suppose that $a+b+c \geq 0$. We wish to show that $A+B+C \geq 0$.

We will show that $A+B+C \geq 0$ by showing that Left wins moving second on $A + B + C$. Any move by Right (on C, say) must be on a negative game. But then, at least one of a, b, and c must also be positive since $a + b + c \geq 0$; without loss of generality $a > 0$. So, Right has moved in the game C to $C + 1$.

---

[7]Note that although we have constructed $\{a \parallel b \mid -c\}$, this does not constitute a proof that TOPPLING DOMINOES is PSPACE-complete, for the construction is exponential in the *number of bits* in a, b, and c.

Then, Left can move on A to $A - 1$. By induction, $(A - 1) + B + (C + 1) \geq 0$. (The induction is valid since the sum of the absolute values of a, b, and c is guaranteed to decrease.) So, after Left's move, the position is a second-player win for Left, and Left thus wins the original game as the second player.

For the converse, suppose that $A + B + C \geq 0$. This means that Left wins moving second from $A + B + C$.

If Right has no move at all, then we have $A, B, C \geq 0$ and $a + b + c \geq 0$ trivially. If, on the other hand, Right has a move available in $A + B + C$, say to $A + B + (C + 1)$, then Left must have a winning countermove. This can be taken to be to the game $(A - 1) + B + (C + 1)$. The sum $(A - 1) + B + (C + 1)$ is a simpler game than $A + B + C$ (since two of the summands are simpler with smaller game trees), and so by induction $(a - 1) + b + (c + 1) \geq 0$. Now, by the associativity and commutativity of integer addition, $a + b + c \geq 0$ as we had hoped.   □

Note that the forward direction in the above induction proof requires no base case, since the proof handles all cases as written. Were you to want to explicitly include $a = b = c = 0$ as a separate case, you would notice that the first player has no move in $A + B + C$ and so loses trivially.

Theorem 5.69. (Theorem 5.6 with proof) $A + B = C$ *if and only if* $a + b = c$.

Proof: From the previous result, $A + B + (-C) = 0$ if and only if $a + b + (-c) = 0$. The stated result now follows by adding C (respectively c) to both sides of each equality.   □

Lemma 5.70. (Lemma 5.16 with proof)

$$a + b + c = 0 \iff A + B + C = 0;$$
$$a + b + c > 0 \iff A + B + C > 0;$$
$$a + b + c < 0 \iff A + B + C < 0.$$

Proof: We will prove these three assertions collectively. Note that since the three conditions are mutually exclusive, we need only prove the forward implication of each of the three assertions, "if $a + b + c = 0$ then $A + B + C = 0$," and so forth, for together they imply the $\iff$.

Suppose that $a + b + c \geq 0$. If Right moves to some $A^R + B + C$, then the definitions of number and integer dictate that $A^R > A$. Hence, $a^R + b + c > 0$, and by induction $A^R + B + C > 0$ and so Right loses. So, we have $A + B + C \geq 0$. A symmetric argument made for $\leq$ gives the first forward implication.

If $a + b + c > 0$, we already argued that Left wins moving second from $A + B + C$, so it suffices to show that Left wins moving first. Now, $a + b + c > 0$ must be a dyadic rational, say $\frac{i}{2^j}$ where either i is odd or $j = 0$. Then, one of the following two cases must hold:

- One of a, b, or c (say a) is of the form $\frac{i'}{2^{j'}}$ for $j' \geq j$ and $j' > 0$, and $a + b + c - \frac{1}{2^{j'}} \geq 0$. In this case, $A^L + B + C \geq 0$ by induction.

- All of a, b, and c are integers, one of which (say a) exceeds 0. Then, $(a - 1) + b + c \geq 0$, and by induction $A^L + B + C \geq 0$.

In both cases, Left has a winning first move, yielding the second forward implication. The third implication is symmetric. □

Note here that there is an implicit induction on the dyadic rationals, a, b, and c. To confirm that the induction terminates, either note that the sum of the denominators of a, b, and c always decreases, or note that the birthdays of A, B, and/or C decrease.

Exercise 5.71. Does the proof properly handle the trivial case when $a = b = c = 0$ as written, or should that case have been included as an easy base case?

Theorem 5.72. (Theorem 5.17 with proof)

- $A + B = C$ if and only if $a + b = c$.

- $A \geq B$ if and only if $a \geq b$.

Proof: The proof is immediate from the lemma and is left as an exercise. □

Exercise 5.73. Prove Theorem 5.17.

Problems

1. What is the value of the PUSH position  ?

2. From a heap of n counters, if $n = 3k$ then both Left and Right can remove one or two counters, if $n = 3k + 1$ then Left can remove one or two counters, and if $n = 3k + 2$ then Right can remove one or two counters. Find the values for all n.

3. From a heap of n counters, if n is even then Left can remove two counters and Right can remove one, and if n is odd then Left can remove one counter and Right can remove two. Find the values for all n.

4. Give a fast way to compute the value of any RED-BLUE CHERRIES path. You need not prove your method correct. Using your method, in a few seconds you should be able to find the value of

⚪⚪⚫⚪⚫⚪⚫⚪⚪⚫⚪⚪⚫⚪⚫⚫⚪⚪

5. Who wins in the following sum of positions from AMAZONS, TOPPLING DOMINOES, and DOMINEERING?

6. Let $g(l, r)$ be the value of EROSION heap (l, r). Use induction to prove

$$g(l, r) = \begin{cases} \lceil \frac{l}{r} - \phi \rceil & \text{if } l \geq r, \\ -\lceil \frac{r}{l} - \phi \rceil & \text{if } l \leq r, \end{cases}$$

where $\phi = \frac{1+\sqrt{5}}{2}$ is the Golden Ratio and satisfies $\phi = \frac{1}{\phi-1}$.

7. Confirm that a number x given by Definition 5.12 is in canonical form.

8. Prove Lemma 5.22 on page 111.

9. Use the result you proved in Problem 5 of Chapter 4 on page 101 to show that LR-HACKENBUSH consists only of numbers.

10. Elwyn Berlekamp found a simple rule for computing the value of an LR-HACKENBUSH string. Assume that the grounded edge is Left's. If all edges in the string are Left's, the value is clearly an integer equal to the number of edges. Otherwise, identify the first left-right alternation. Left's edges before the alternation contribute 1 each. Replace the two alternating edges by a decimal point and replace each subsequent left (respectively, right) edge by a 1 (respectively, 0) and append a 1. You can now read off the fractional value in binary. For example,

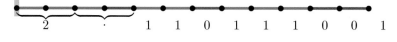

$$= 2 + .110111001 = 2 + \frac{1}{2} + \frac{1}{4} + \frac{1}{16} + \frac{1}{32} + \frac{1}{64} + \frac{1}{512} = 2\frac{441}{512}.$$

Thea van Roode has another way of assigning values to Hackenbush strings. Assign value 1 to edges until the first color change. Thereafter, divide by 2 at each new edge, and the sign of each edge depends on its color. For example,

$$\begin{array}{ccccccccccc} 1 & 1 & 1 & -\frac{1}{2} & \frac{1}{4} & \frac{1}{8} & -\frac{1}{16} & \frac{1}{32} & \frac{1}{64} & \frac{1}{128} & -\frac{1}{256} & -\frac{1}{512} \end{array}$$

$$= 1 + 1 + 1 - \frac{1}{2} + \frac{1}{4} + \frac{1}{8} - \frac{1}{16} + \frac{1}{32} + \frac{1}{64} + \frac{1}{128} - \frac{1}{256} - \frac{1}{512} = 2\frac{441}{512}.$$

Prove that either (or both) of these methods work.

11. (a) Call f_n the value of the following HACKENBUSH position as a function of n, the number of legs. Determine f_n for small values of n. How far can you go?

(b) How about this position?

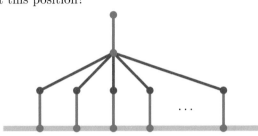

(c) Come up with an infinite series of HACKENBUSH positions of your own. See if you can find the first few values.

12. Determine the left and right incentives from TOPPLING DOMINOES positions of the form

$$\mathbf{1}^m \mathbf{1}^n$$

Give specific guidance on how you should play on sums and differences of these games. (This is a follow-up to Problem 7 of Chapter 4.)

13. For each of the following MAZE positions, confirm the value of the starting square at the top by computing the canonical form of all start positions, working your way up as you go.

(a) The MAZE position in Exercise 2.10 on page 41 has value $\frac{1}{4}$.

(b) The two mazes in Problem 1 of Chapter 2 on page 54 have values 0 and $\{1 \mid -\frac{1}{2}\}$, respectively.

(c) The following two mazes have values $\{0, *, \{0, * \mid -1\} \mid -1\}$ and $\{-\frac{1}{2}, \{-\frac{1}{2} \mid -1\} \mid -1\}$, respectively:

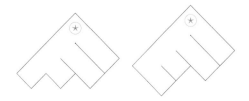

14. Prove that any all-small is infinitesimal. That is, if G is all-small, then $-x < G < x$ for all positive numbers x.

15. In this problem, abbreviate PUSH positions using superscripts for repetition of blank squares. So, 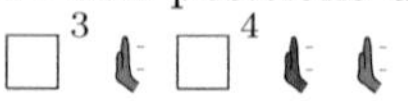 is the position

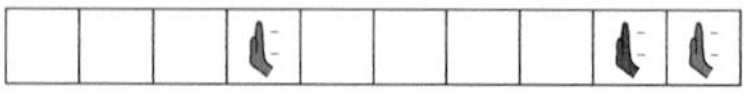

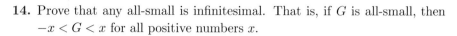

Prove the following:

(a) 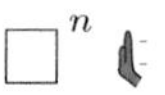 has value $n+1$.

(b) 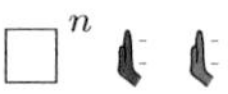 has value $2 - \frac{1}{2^{n+1}}$.

(c) 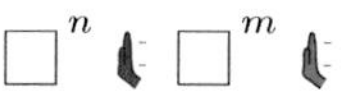 (where $m > 0$) has value $m+1$.

16. Prove Theorem 5.27 on page 112.

17. Prove that if G either has no right options or has no left options, then G is an integer.

18. Prove Theorem 5.48, which gives the canonical form of $n{\cdot}\uparrow$ and $n{\cdot}\uparrow*$. You should use diagrams (like those preceding the statement of the theorem) to indicate reversible and dominated options.

19. Let $g(a, b, c)$ be the one-dimensional AMAZONS position with a blank squares, then a blue amazon, then b blank squares, then a red amazon, and then c blank squares. For example,

$g(5, 2, 3) =$ 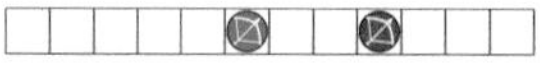

What is the value of $g(a, b, c)$?

20. Define $g(a, b, c, d)$ similar to the last problem but with one additional blue amazon and one additional gap. For example,

$g(5, 2, 3, 1) =$ 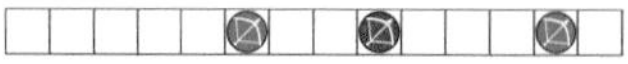

What is the value of $g(a, b, c, d)$? (Be sure to check whether and when special case(s) are required if $abcd = 0$.)

21. One expects $g+g$ to have a more appealing canonical form than $g+g+g$. This problem explores a counterexample. In particular, let $x > 0$ be a number; then $+_x$ $_x$ $_x$ can be rewritten as $_G$, where G is infinitesimally close to x.

(a) Compare $_x$ and $\{0 \mid -x\}$.

(b) Compare $_x$ $_x$ and $\{0 \mid -x\}$.

(c) Compute the canonical form of $\uparrow_x *_x$. (One position should have either multiple left options or multiple right options.)

(d) Compute the canonical form of $*_x *_x *_x$, and write the result in the form $*_G$.

Determine G by hand, showing all your work. (You may use software to help you find mistakes as you work.)

22. John Conway [Con01, p. 215] observes that, "It is amusing to verify that for *any* game G, we have $*_{*_G} = \uparrow$, so that, in particular, $\uparrow$ is the unique solution of $G = *_G$." Prove all of these observations.

23. For integers a and c, what is the canonical form of $g = \{a \,\|\, 0 \,|\, -c\}$? Naturally, you will need several cases depending, in part, on the order of a, 0, and c. If a game (or a position of the game) has a value that appeared in this chapter, be sure to name it.

Preparation for Chapter 6

Prep Problem 6.1. One can define the partial order on positive integers where $a \leq b$ if a exactly divides b. Here is a diagram (called a Hasse diagram) of the factors of 12 under this partial order:

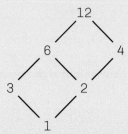

Note that in the diagram, a line is drawn from a upward toward b if $a < b$ and no other c fits in between (i.e., any comparisons can be inferred from the diagram).

(a) Draw two Hasse diagrams, one each for the factors of 36 and of 30.

(b) How would you determine the greatest common divisor (gcd) and the least common multiple (lcm) of two elements by looking at Hasse diagrams?

Prep Problem 6.2.

(a) List all games each with a single left option and a single right option each chosen from $\{1, 0, -1\}$. Draw the partial order of these nine games. (The diagram should look like that of the factors of 36 from Prep Problem 6.1.)

(b) How many additional games would there be if we allowed the left or right options to be empty? List them.

To the instructor: While we make use of the material in Section 6.3 in Chapter 8, the material in Sections 6.4 and 6.5 can be skipped without loss of continuity.

Chapter 6

Structure

... a chess player may offer the sacrifice of a pawn or even a piece, but a mathematician offers the game.

Godfrey Hardy in *A Mathematician's Apology*

In this chapter we will prove theorems that say something about all games, or about games born on or before a given day. The sorts of questions that we hope to answer are the following:

- In what ways can we classify games in an informative or interesting way?

- What are some extremal games: biggest, smallest, etc.?

- We know that games form a group with a partial order. Are there other underlying algebraic structures in classes of games?

6.1   Games Born by Day 2

One way of classifying games is by their birthday. Recall from Definition 4.1 on page 76 that the birthday of a game is the height of its game tree. A game is *born by* day n if its birthday is less than or equal to n. Define

$$\mathcal{G}_n = \text{the set of games born by day } n,$$

and let
$$g_n = |\mathcal{G}_n| = \text{the number of games born by day } n.$$

There is only one game born on day 0, and that is the game 0. So, in our notation, $\mathcal{G}_0 = \{0\}$ and $g_0 = 1$. There are $g_1 = 4$ games born by day 1, those being $\mathcal{G}_1 = \{1, *, 0, -1\}$. The left and right options of games from $\mathcal{G}_n$ are subsets of $\mathcal{G}_{n-1}$. Since the number of subsets of a set of n elements is 2^n, we have the following observation.

Observation 6.1. $g_n \leq 2^{g_{n-1}} \cdot 2^{g_{n-1}} = 2^{2g_{n-1}}$.

Many of the games constructed in this way will not be in canonical form, so we expect that the actual value of g_n will be much less than that provided by this estimate. On day 2, the observation states that there are at most $2^8 = 256$ games. But, note that if two comparable options are available to a player, one of the two will be dominated. Consequently, candidate sets of left or right options are *antichains* of games born by day $n - 1$. An *antichain* is a set consisting only of incomparable elements. There are six antichains of games born by day 1:

$$\{1\}, \{0, *\}, \{0\}, \{*\}, \{-1\}, \{\,\}.$$

That reduces the potential number of games born by day 2 to 36, as shown in the following table. These six antichains are arranged, in some sense, with Left's preferred left options listed first. (We have also dropped the brackets.) In particular, Left wins moving first if Left's options are any of the first three. The same six antichains are sorted for Right in a similar fashion.

Right

	-1	$0, *$	0	$*$	1	$\emptyset$
1						
$0, *$		$\mathcal{N}$			$\mathcal{L}$	
0						
$*$						
-1		$\mathcal{R}$			$\mathcal{P}$	
$\emptyset$						

(Left labels the rows)

The lower-right quadrant consists entirely of games equal to 0. Three other pairs of positive positions (and, symmetrically, negative positions) also turn out to be equal, so it turns out that only 22 games are born by day 2:

Right

	-1	$0, *$	0	$*$	1	$\emptyset$
1	± 1	$1\|0, *$	$1\|0$	$1\|*$	$1*$	2
$0, *$	$0,*\|-1$	$*2$	$\uparrow*$	$\uparrow$	$\frac{1}{2}$	1
0	$0\|-1$	$\downarrow*$	$*$			
$*$	$*\|-1$	$\downarrow$			0	
-1	$-1*$	$-\frac{1}{2}$				
$\emptyset$	-2	-1				

(Left labels the rows)

Exercise 6.2. Confirm that the three non-canonical positive values, $\{0, * \mid *\}$, $\{0, * \mid 1\}$, and $\{0, * \mid \}$, are equal to their purported values, $\uparrow$, $\frac{1}{2}$, and 1, respectively, by converting each to canonical form. Clearly identify any dominated and/or reversible options.

We can also investigate the partial order of just those games born by day n. Figure 6.1 shows *Hasse diagrams* of the partial orders of games born by day 1 and by day 2. In a Hasse diagram, two games $G > H$ are joined by a line if one game is greater than the other, but for no J is $G > J > H$. The greater game is always above the lesser one. For example, $\uparrow * > *2$, and they are joined by a line because there is no other game born by day 2 whose value lies between them. However, $\uparrow > \downarrow *$, but they are not joined by a line because $\uparrow > *2 > \downarrow *$. Since there is no monotonic path from $\downarrow$ to ± 1, the diagram shows that these two games are incomparable.

There are exactly 1474 games born by day 3. Using bounds from [WF04], the number of games born by day 4 is somewhere between 3 trillion (3×10^{12}) and 10^{434}.

6.2 Extremal Games Born by Day n

In this section, we will describe the greatest and least positive games and the greatest infinitesimals born by day n.

Theorem 6.3. *The greatest game born by day n is n.*

Proof: We wish to show that Left wins moving second from $n - G$ whenever G is born by day n. A winning strategy for Left is simply to move on n for n consecutive turns. The birthday of a game is one more than the length of the longest branch in its game tree, so since Right's only legal moves are on G and G is born by day n, Right will run out of moves before Left does.   □

An inductive proof, while neither more revealing nor more clear, may help to motivate other proofs in this section.

Alternate Proof of Theorem 6.3: We wish to show that Left wins moving second from $n - G$ whenever G is born by day n. Right's only legal moves are to some $n - G^L$, and Left can respond to $(n - 1) - G^L$. By the definition of birthday, G^L must be born by day $n - 1$, and so by induction Left wins moving second on $(n - 1) - G^L$.   □

Theorem 6.4. *The least positive number born by day $n + 1$ is 2^{-n}.*

Proof: Suppose that $x > 0$ is a number born by day $n + 1$. Without loss of generality, x is in canonical form, for conversion to canonical form can only

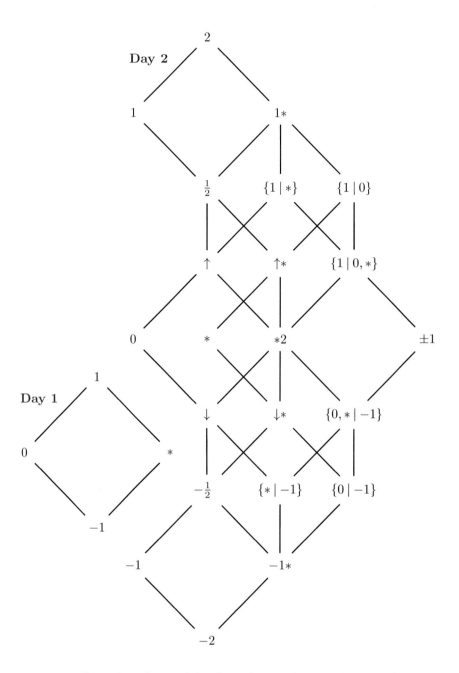

Figure 6.1. The partial orders of games born by days 1 and 2.

reduce birthday. So, x is of the form $\{y \mid z\}$ for $0 \le y < z$. For x to be minimal, y and z are chosen as small as possible; i.e., $y = 0$ and z is the minimum positive number born by day n, which by induction is 2^{1-n}. Since $\{0 \mid 2^{1-n}\} = 2^{-n}$, we have that 2^{-n} is the least positive number born by day $n + 1$.

For the base case of $n = 0$, $1 = 2^0$ is the only positive game born on day 1. $\square$

Theorem 6.5. *The least positive game born by day $n + 2$ is $\boldsymbol{+}_n$.*

Proof: Let $G > 0$ be born by day $n+2$. We wish to show that Left wins moving second on $G - \boldsymbol{+}_n = G + \boldsymbol{-}_n = G + \{n \mid 0 \parallel 0\}$:

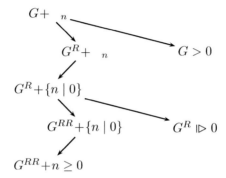

If Right moves the second component to 0, Left wins since G is positive. If, on the other hand, Right starts on G, Left responds on the second component, leaving $G^R + \{n \mid 0\}$. From there, if Right moves the second component to 0, then Left has a winning move from G^R since G was positive. If, on the other hand, Right moves on the first component, Left moves on the second, leaving $G^{RR} + n$. Since G was born by day $n + 2$, G^{RR} was born by day n, and hence, by Theorem 6.3, Left wins $G^{RR} + n$ moving second. $\square$

The last theorem along these lines is a bit more complicated. There are two maximal infinitesimals born by day $n + 1$, those being $n \cdot \uparrow$ and $n \cdot \uparrow *$. For example, Figure 6.2 shows just the infinitesimals born by day 2.

Theorem 6.6. *For any infinitesimal G born by day $n + 1$, either $G \le n \cdot \uparrow$ or $G \le n \cdot \uparrow *$.*

Proof: Lemma 6.9 on page 145 generalizes this result. $\square$

To motivate additional machinery required to prove this theorem, we will first prove a special case of the theorem that applies only to all-small games.

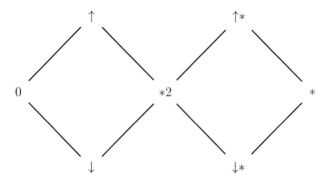

Figure 6.2. The partial order of infinitesimals born by day 2.

Lemma 6.7. *For any all-small game G born by day $n + 1$, either $G \leq n{\cdot}\uparrow$ or $G \leq n{\cdot}\uparrow{*}$.*

Proof: Assume that $n \geq 2$ and let G be an all-small game born by day $n + 1$. We wish to show that Left wins moving second on $n{\cdot}\uparrow - G$ or on $n{\cdot}\uparrow{*} - G$. Suppose that both games are played simultaneously, and Right moves first in both. To prove the theorem, Left need only win one of the two games.

Suppose that Right moves $-G$ to $-G^L$ in either game. If G^L is a number, it must be 0, and the resulting difference game is positive. If G^L is not a number, Left replies by moving to $-G^{LR}$, which is born by day $n - 1$. Left then wins by induction, for both $n{\cdot}\uparrow$ and $n{\cdot}\uparrow{*}$ exceed both of the games $(n - 2){\cdot}\uparrow$ and $(n - 2){\cdot}\uparrow{*}$.

Suppose that Right moves to the pair of games $(n-1)\uparrow{*}-G$ and $(n-1)\uparrow-G$. Then, Left can move both copies of $-G$ to the same $-G^R$. Left, moving second, can proceed to win either $(n - 1){\cdot}\uparrow{*} - G^R$ or $(n - 1){\cdot}\uparrow - G^R$ by induction.

The base cases are when $n = 0$ and $n = 1$. On day 1, the games 0 and $*$ are the only infinitesimals. Day 2 can be confirmed by verifying that Figure 6.2 is correct.   □

The reason that the same proof fails to prove Theorem 6.6 is that in the two places induction is used, we have no guarantee that G^{LR} and, respectively, G^R are infinitesimal games. For example, if $G = \{\mathbf{+}_2 \mid 0\}$, then $G^{LR} = \{0 \mid -2\}$ is not infinitesimal. Similarly, if $G = \mathbf{+}_2$, $G^R = \{0 \mid -2\}$, which is not infinitesimal.

Exercise 6.8. Take a moment to make sure that you understand and appreciate the first sentence of the preceding paragraph.

In both of these examples, the appearance of negative numbers in G (meaning positive numbers in $-G$) should make it *easier* for Left to win $n{\cdot}\uparrow - G$.

Put differently, in the induction argument, Right is never given the opportunity to move twice in a row in $-G$ and should therefore never be able to reach a negative number in $-G$, for that would contradict the fact that G is infinitesimal. Left, however, may move twice in a row, so while she may reach a positive number in $-G$, that fails to hurt the argument since Left wins even more decisively.

To formalize the line of reasoning in the last paragraph, we require the notion of left and right stops from Section 5.3. We are now ready to prove Theorem 6.6. Recall from the discussion following Lemma 6.7 that the challenge is in accounting for the possibility that G^{LR} and G^R might not be infinitesimal, making the induction argument fail. So, we will strengthen the induction hypothesis to allow G to take on some non-infinitesimal values:

Lemma 6.9. *If G is born by day $n + 1$ with $\mathbf{LS}(G) \leq 0$, then either $G \leq n \cdot \uparrow$ or $G \leq n \cdot \uparrow *$.*

Theorem 6.6 is an immediate corollary since all infinitesimals have zero stops.

Proof: Fix $n \geq 2$, and fix G with $\mathbf{LS}(G) \leq 0$ that is born by day $n + 1$. We wish to show that Left wins moving second on $n \cdot \uparrow - G$ or on $n \cdot \uparrow * - G$. Suppose that both games are played simultaneously and that Right moves first in both. We will show that Left can win one of the two games.

If G is already a number, $G \leq 0$ and the result is trivial.

Suppose that Right moves $-G$ to $-G^L$ in either game. If G^L is a number, it must be less than or equal to 0, and the resulting difference game is positive. If G^L is not a number, since $\mathbf{LS}(G) \leq 0$ Left can reply locally to some $-G^{LR}$ (which is born on day $n - 1$) with $\mathbf{LS}(G^{LR}) \leq \mathbf{LS}(G) \leq 0$. Left then wins by induction, for $n \cdot \uparrow$ and $n \cdot \uparrow *$ exceed both of the games $(n - 2) \cdot \uparrow$ and $(n - 2) \cdot \uparrow *$.

Suppose that Right moves to the pair of games $(n-1)\uparrow * - G$ and $(n-1)\uparrow - G$. Then, Left can move both copies of $-G$ to the same $-G^R$, where G^R is chosen with minimal left stop. Then, $\mathbf{RS}(G^R) \leq \mathbf{LS}(G^R) = \mathbf{RS}(G) \leq \mathbf{LS}(G) \leq 0$. Left, moving second, can proceed to win either $(n-1)\cdot\uparrow * - G^R$ or $(n-1)\cdot\uparrow - G^R$ by induction.

The base cases of $n = 1$ and $n = 2$ can be confirmed by checking Figure 6.1. □

Exercise 6.10. We tacitly assumed that if $\mathbf{LS}(G) \leq 0$ and G^L is not a number, then some G^{LR} has $\mathbf{LS}(G^{LR}) \leq \mathbf{LS}(G)$. Confirm this fact.

Exercise 6.11. Justify each of the relations appearing in the second-to-last paragraph of the proof, those being

$$\mathbf{RS}(G^R) \leq \mathbf{LS}(G^R) = \mathbf{RS}(G) \leq \mathbf{LS}(G) \leq 0.$$

6.3   More About Numbers

We are now prepared to prove a few more theorems about numbers. To review, we know from page 109 that, in canonical form, numbers have negative incentives. The Weak Number-Avoidance Theorem states that if x is a number and G is not, and if Left can win $x + G$ moving first, then she can do so with a move on G. We now state the strong version of this theorem:

Theorem 6.12. (Number Avoidance) *Suppose that x is a number in canonical form with a left option and that G is not a number. Then, there exists a G^L such that $G^L + x > G + x^L$.*

Proof: Equivalently, we wish to show that some $G^L - G$ exceeds $x^L - x$. It is clearly sufficient to show that some left incentive exceeds all negative numbers. In order to establish this, it suffices to prove that for some G^L, $\mathbf{RS}(G^L - G) \geq 0$, for then Lemma 6.9 tells us that $G^L - G \geq n \cdot \downarrow$ or $G^L - G \geq n \cdot \downarrow *$ for some value of n, where $n + 1$ is G's birthday. This implies that $G^L - G$ exceeds all negative numbers.

   To see that for, some G^L, $\mathbf{RS}(G^L - G) \geq 0$, choose G^L to be a left option with maximum right stop. Then, $\mathbf{RS}(G^L) = \mathbf{LS}(G)$. Left has a strategy playing second on $G^L - G$ to achieve a non-negative stopping value by responding locally to each of Right's moves, achieving at least $\mathbf{RS}(G^L) + \mathbf{RS}(-G) = \mathbf{RS}(G^L) - \mathbf{LS}(G) = 0$. If Right's move reaches a number in one component, Left can move twice in a row on the other, in which case Theorem 5.34 says that she could do no worse.

   There is one subtlety — the difference of the two components might be a number (yielding a stop). Allow both players to play on as above except allow Right to pass, which (since optional) can only help Right. If, in fact, the difference is a number (having negative incentive), Left's consecutive moves can only hurt her. Despite this change, the above argument still leaves Left at a belated stop that is at least 0.   □

Theorem 6.13. (Number-Translation) *If x is a number and G is not, then $G + x = \{\mathcal{G}^L + x \mid \mathcal{G}^R + x\}$.*

Proof: $G+x = \{\mathcal{G}^L + x, G + x^L \mid \mathcal{G}^R + x, G + x^R\}$. By the Number-Avoidance Theorem, the last option on each side, if it exists, is dominated by one of the previous options.   □

   The Number-Translation Theorem works both ways — adding a number into a game and taking one out of all the options.

Example 6.14. How should the following disjunctive sum be played:

$$G = \{5 \mid 0\} + \{-2 \mid -7\} + \{3 \mid -1\} + \{2 \mid -2\} + \{1 \mid 0\}?$$

The second and third summands can be re-written to give

$$G = \{5 \mid 0\} + (-2 + \{0 \mid -5\}) + (1 + \{2 \mid -2\}) + \{2 \mid -2\} + \{1 \mid 0\}$$
$$= (-2 + 1) + (\{5 \mid 0\} + \{0 \mid -5\}) + (\{2 \mid -2\} + \{2 \mid -2\}) + \{1 \mid 0\}$$
$$= -1 + 0 + 0 + \{1 \mid 0\}$$
$$= \{0 \mid -1\}.$$

Since $G = \{0 \mid -1\}$, both players can win moving first. A simple way to win is for the first player to move in $\{1 \mid 0\}$, and then $\{5 \mid 0\}$ and $\{-2 \mid -7\}$ are paired off as good responses to each other, and so are $\{3 \mid -1\}$ and $\{2 \mid -2\}$.

Theorem 6.15. (Negative Incentives) *If all of G's incentives are negative, then G is a number.*

Proof: Suppose, to the contrary, that G is not a number and all its incentives are negative.

If $G \parallel x$, where x is a number, then Left wins moving first on $G - x$ and (by Number-Translation) some $G^L - x \geq 0$. But then, since G's incentives are negative, we have $G - x > G^L - x \geq 0$, contradicting $G \parallel x$. So, G is comparable with all numbers; thus, $\mathbf{LS}(G) = \mathbf{RS}(G)$ and G must be a number plus an infinitesimal comparable with 0, say greater than 0. If G's incentives are negative, so are those of $G - x$ and $-G$. We can translate our counterexample to make the number portion 0.

So, without loss of generality, G is a positive infinitesimal.

Next, we will show (by induction) that if s is all-small, then $G + s > 0$. Since $G + s^L$ and $G + s^R$ are positive (by induction), we only need to confirm that Left wins moving second if Right plays to some $G^R + s$. Left responds to any $G^R + s^L$, and we have $G^R + s^L > G + s^L > 0$. (The first inequality is because G's incentives are negative, and the second is by induction.)

So, $G + n \cdot \downarrow > 0$ for all positive integers n, but Theorem 6.6 then provides a contradiction, since no infinitesimal has this property. □

Exercise 6.16. The preceding proof concludes from $\mathbf{LS}(G) = \mathbf{RS}(G)$ that G must be a number plus an infinitesimal. Use Theorem 6.13 and the result of Problem 3 to prove this statement.

The stops give another common way of classifying games into categories:

Definition 6.17. A game is dubbed *cold* if it is a number, *tepid* if $\mathbf{LS}(G) = \mathbf{RS}(G)$ but G is not a number, and *hot* if $\mathbf{LS}(G) > \mathbf{RS}(G)$.

Observation 6.18. Every game is either cold, tepid, or hot. Every tepid game is a number plus a non-zero infinitesimal.

6.4   The Distributive Lattice of Games Born by Day n

We can investigate the partial order of games (how they compare) as distinct from the group of games (how they add). Refer back to Figure 6.1 on page 142, which shows the partial order of games born by days 1 and 2. This is not merely *any* partial order. It has quite a bit of structure.

Definition 6.19. A *lattice* is a partial order with the additional property that any pair of elements a and b in the partial order has a *least upper bound* or *join,* denoted $a \vee b$, and a *greatest lower bound* or *meet,* denoted $a \wedge b$.

For example, the following partial order is not a lattice for a number of reasons:

For one, a and b have no join, since the only two elements greater than or equal to both, c and d, are incomparable. So, a and b have no *least* upper bound. Secondly, there are no elements greater than or equal to both c and d, so c and d have no *upper bound* at all, let alone a least upper bound.

Common lattices in mathematics include the integers under the ordinary $\geq$ and the positive integers where we define $b \geq a$ if a is a divisor of b.

Theorem 6.20. *The games born by day n form a lattice.*

Before embarking on the proof, it is worthwhile to explore the theorem's meaning.

Exercise 6.21. Referring to Figure 6.1, identify the joins and meets of the following pairs of elements among the games born by day 2:

(a) $* \vee 0$,

(b) $* \vee \frac{1}{2}$,

(c) $\uparrow \vee \pm 1$,

(d) $\uparrow \wedge \pm 1$.

Recompute your answers using the definitions of $\vee$ and $\wedge$ given in the proof below.

Proof: We will prove Theorem 6.20 by explicitly constructing the join operation (and, symmetrically, the meet) and then confirming that it is, indeed, a least upper bound (respectively, greatest lower bound). Recall that $\mathcal{G}_k$ denotes the set of games born by day k. For $G \in \mathcal{G}_n$ define

$$\lceil G \rceil \overset{\text{def}}{=} \{H \in \mathcal{G}_{n-1} \mid H \rhd G\};$$

$$\lfloor G \rfloor \overset{\text{def}}{=} \{H \in \mathcal{G}_{n-1} \mid H \lhd G\}.$$

Now let

$$G_1 \vee G_2 \overset{\text{def}}{=} \left\{ G_1^L \cup G_2^L \mid \lceil G_1 \rceil \cap \lceil G_2 \rceil \right\};$$

$$G_1 \wedge G_2 \overset{\text{def}}{=} \left\{ \lfloor G_1 \rfloor \cap \lfloor G_2 \rfloor \mid G_1^R \cup G_2^R \right\}.$$

Since all the games that are options of $G_1 \vee G_2$ or $G_1 \wedge G_2$ belong to $\mathcal{G}_{n-1}$, the games $G_1 \vee G_2$ and $G_1 \wedge G_2$ belong to $\mathcal{G}_n$. We wish to show further that $G_1 \vee G_2$ is a least upper bound for G_1 and G_2 in $\mathcal{G}_n$. To that end, we need to show that

- $G_1 \vee G_2 \geq G_i$ (for $i = 1, 2$), and

- if $G \in \mathcal{G}_n$ and $G \geq G_1$ and $G \geq G_2$, then $G \geq G_1 \vee G_2$.

Referring to the definition of join above, we will prove each of these two assertions in sequence using diagrams.

- Left wins the following game moving second:

$$
\begin{array}{c}
(G_1 \vee G_2) - G_i \\
\diagdown \qquad\qquad\qquad\nearrow \\
(G_1 \vee G_2) - G_i^L \qquad\qquad H - G_i \\
\diagup \\
G_i^L - G_i^L
\end{array}
$$

where (in the second right option) $H \in \lceil G_1 \rceil \cap \lceil G_2 \rceil$. In particular, $H \rhd G_i$, and so Left wins moving first from $H - G_i$.

- If $G \geq G_1$ and $G \geq G_2$, then Left wins moving second from

$$
\begin{array}{c}
G - (G_1 \vee G_2) \\
\diagdown \qquad\qquad\qquad\nearrow \\
G^R - (G_1 \vee G_2) \qquad\qquad G - G_i^L \\
\diagup \\
G^R - G^R
\end{array}
$$

In the first case, observe that, since $G - G_i \geq 0$, we must have $G^R - G_i \rhd$ 0, i.e., $G^R \quad G_i$ for $i = 1, 2$. Furthermore $G^R \in \mathcal{G}_{n-1}$. Together these imply that $G^R \in \lceil G_1 \rceil \cap \lceil G_2 \rceil$, which means that the option indicated for Left from $G^R - (G_1 \vee G_2)$ exists. In the second case, since $G \geq G_i$, we also have $G - G_i^L \quad 0$, so Left will win. $\qquad\square$

Definition 6.22. A *distributive lattice* is a lattice in which a meet distributes over a join (or, equivalently, a join distributes over a meet). That is,

$$a \wedge (b \vee c) = (a \wedge b) \vee (a \wedge c).$$

The parenthetical remark is not intended to be obvious, but it has a short proof easily found in textbooks covering lattice theory; see, for example, [Grä71, p. 35] or [CD73, p. 19]. Not all lattices are distributive; the lattices in the following two diagrams are not:

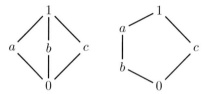

Exercise 6.23. Compute $a \wedge (b \vee c)$ and $(a \wedge b) \vee (a \wedge c)$ for each of the above lattices to confirm that they are not distributive.

In fact, in some sense, these are the *only* non-distributive lattices. Garrett Birkhoff showed that a lattice is distributive *if and only if* it fails to contain both of the above as a sublattice.

Theorem 6.24. *The games born by day n form a distributive lattice.*

Proof: See Problem 5. $\qquad\square$

6.5 Group Structure

In the last section, we investigated the partial order structure of games born by day n, as distinct from the group structure. That is, we focused on the $\geq$ symbol rather than the $+$ sign. We can also do the reverse by looking at the group generated by games born by day n. In other words, what are the possible sums of games born by day n?

On day 0, only the game 0 is born. Not much to do there.

On day 1, we get 4 games, $\{1, *, 0, -1\}$. Since $*$ is of order 2 (meaning $* + * = 0$), sums of games born by day 1 are either an integer or an integer plus *. In group-speak, the group generated by games born by day 1 is just $\mathcal{Z} \times \mathcal{Z}_2$. Here, $\mathcal{Z}$ denotes the group of integers under addition and $\mathcal{Z}_2$ the group of integers modulo 2 under addition (more generally, $\mathcal{Z}_k$ denotes the group of integers modulo k under addition).

In [Moe91], David Moews does a similar analysis on all games born by day 2 and on games born by day 3, ignoring infinitesimal shifts. For example, the group of games generated by those born by day 2 has an independent generating set:

$$\frac{1}{2}, *2, A, \uparrow, \alpha, \pm\frac{1}{2}, \pm 1,$$

where

$$A = \{1 \mid 0\} - \{1 \mid *\},$$
$$\alpha = \{1 \mid 0\} - \{1 \mid 0, *\} = \{1* \mid * \,\|\, 0 \mid -1\}.$$

Note that A has order 4 since $A + A = *$, while $\alpha > 0$ but $n \cdot \alpha \lhd \uparrow$ for any $n > 0$ and is therefore additively independent of $\uparrow$. So, the group of games generated by those born by day 2 is isomorphic to $\mathcal{Z}^3 \times \mathcal{Z}_4 \times \mathcal{Z}_2^3$.

Exercise 6.25. Confirm that $n \cdot \alpha \quad \uparrow$ for all $n > 0$.

Problems

1. Curiously, each of the 22 games born by day 2 appears as a KONANE position with four stones, two of each color, on (say) an 8×8 board. Demonstrate that this is possible by constructing each position (or its negative).

2. How many statements of the form $G > H$ and $G \parallel H$ must be confirmed to check that Figure 6.1 (day 2 diagram) is accurate? As you count, try to exploit symmetries or other space-saving techniques. Naturally, you should explain your higher-level reasoning used to shorten your list. (For example, if $G > H$, then $-H > -G$, so there is no need to explicitly confirm both.)

   Do not prove each of the statements; that would get old very fast.

3. Prove that G is infinitesimal *if and only if* its stops are 0.

4. (a) Show that the sum of a tepid game and a hot game is hot.

   (b) Show that the sum of two hot games can be tepid or even cold.

5. Prove Theorem 6.24 by working through the following exercises. We wish to show that $S_1 = S_2$, where

$$S_1 = H \wedge (G_1 \vee G_2),$$
$$S_2 = (H \wedge G_1) \vee (H \wedge G_2).$$

(a) Confirm that

$$\lfloor G_1 \vee G_2 \rfloor = \lfloor G_1 \rfloor \cup \lfloor G_2 \rfloor,$$
$$\lceil G_1 \wedge G_2 \rceil = \lceil G_1 \rceil \cup \lceil G_2 \rceil.$$

(b) Show that

$$S_1 = \{\, \lfloor H \rfloor \cap \lfloor G_1 \vee G_2 \rfloor \mid H^R, \lceil G_1 \rceil \cap \lceil G_2 \rceil \,\},$$
$$S_2 = \{\, \lfloor H \rfloor \cap \lfloor G_1 \vee G_2 \rfloor \mid \lceil H \rceil, \lceil G_1 \rceil \cap \lceil G_2 \rceil \,\}.$$

(c) Prove that $S_1 = S_2$ by playing $S_1 - S_2$.

6. Consider the following conjecture: If G is in canonical form, all its left stops (not just the maximum one) exceed all of its right stops. More precisely, if G is canonical and not a number, then for every G^L and G^R, $\mathbf{RS}(G^L) \geq \mathbf{LS}(G^R)$.

Prove the conjecture false by providing a counterexample.

7. What are the left and right stops for CLOBBER THE POND positions ⬤ ⬤⬤ and ⬤ ⬤ ⬤⬤ ?

8. In the paper, "Snakes in Domineering Games," one of the authors (Wolfe) claimed that

While the fact may be true, it remains conjecture.

Explain what is wrong with the following "proof": We wish to show that the second player wins the difference game

If either Left or Right plays entirely in one of the G or H pieces, the second player can respond in the other, leaving (by induction) 0. If Left moves covering the squares adjacent to **b** or **c**, Right responds at **x**, leaving a position ≤ 0. If Left moves covering the squares adjacent to **a**, then Right responds at **d**, leaving 0 since

(This latter statement is an application of Problem 6 of Chapter 4 on page 101.) If Right moves covering the squares adjacent to **d** or **e**, then Left responds at **a** or **b** (respectively), leaving a game ≥ 0 by the One-Hand-Tied Principle. Lastly, Right's move at **x** need not be considered by the Number-Avoidance Theorem.

Preparation for Chapter 7

Prep Problem 7.1. Review binary (base 2) arithmetic. If you need a refresher, type "base 2" into wikipedia.org. Complete the missing entries in the following table:

Decimal fractional notation	Binary positional notation
$-2\frac{1}{2}$	-10.1
23	
	10101
$\frac{1}{4}$	
$\frac{5}{8}$	
	$.01101$
$-12\frac{9}{16}$	
	-11.10101

(In truth, you will not need to deal with binary fractions in this chapter, but they appear frequently in combinatorial game theory. See, for example, Problem 10 of Chapter 5 on page 134.)

To the instructor: GREEN HACKENBUSH from [BCG01, pp. 189–196] is completely solved, and the proofs are just hard enough to be a nice challenge for the solid undergraduate student. You could cover the material immediately after Section 7.3.

Chapter 7

Impartial Games

The real excitement is playing the game.

Donald Trump

If, for all the positions of a game, the left options equal the right options, then the game is dubbed *impartial*. The rulesets for legal moves in impartial games make no distinction between the players. For example, in a popular impartial game among children, there is a heap of n counters, and a move consists of removing one, two, or three counters from the heap. We have also seen the impartial variant of DOMINEERING known as CRAM in Chapter 0.

One important feature of an impartial game is that, at least intuitively, neither Left nor Right can accumulate an advantage, since if Left has a legal move then Right also has that move available.[1] In addition, recall Theorem 2.13 from page 43, that any impartial game is either an $\mathcal{N}$-position or a $\mathcal{P}$-position.

Exercise 7.1. Every impartial game is its own negative. That is, if G is impartial, then $G + G = 0$.

By definition the impartial games are a subset of the all-small games, and therefore the value of an impartial game will be an infinitesimal. However, the set of values that can occur for impartial games is a very limited subset even of the infinitesimals. For instance, no impartial game can have value $\uparrow$ since $\uparrow > 0$; that is, $\uparrow \in \mathcal{L}$. Also, while $\uparrow* \in \mathcal{N}$, $\uparrow* + \uparrow* = \Uparrow \neq 0$. So, by the previous exercise, no impartial game can have value $\uparrow*$. An alternative argument would be to consider the canonical form of $\uparrow$, which is $\{0 \mid *\}$. This has different left and right options and is therefore not impartial.

Exercise 7.2. Confirm the implicit claim made in the last paragraph that the *canonical form* of an impartial game has the same left and right options.

---

[1] The reader experienced with playing NIMSTRING might reasonably question this assertion.

Finding the value of an impartial game is typically much easier than finding the value of an arbitrary partizan game. Because both left and right options are the same, we only have to list, say, Left's options. This also allows us to introduce later, in Section 7.4, a more concise notation.

When an impartial game does not naturally decompose into a disjunctive sum, it is usually easiest and most useful to restrict attention to outcome classes alone, as we did in Section 2.3. However, when the game does naturally decompose, identifying values is usually the more powerful method for describing and analyzing the game.

NIM was the first combinatorial game to have its full strategy published in 1901 [Bou02]. In the 1930s, 1940s, and 1950s Roland Sprague [Spr35], Patrick Grundy [Gru39], and Richard Guy [GS56] showed that the theory for NIM could be extended to all impartial games and analyzed several games using the theory. We follow this development, but with a modern treatment. A playful and thorough treatment of impartial games is to be found in Richard Guy's lovely paperback *Fair Game: How to Play Impartial Combinatorial Games* [Guy89].

7.1   A Star-Studded Game

There is a class of infinitesimals that do not behave like any values that we have yet encountered. They occur naturally as the values of the impartial game NIM, and as we shall see, they are the only values that impartial games take on!

NIM is played with heaps of counters (or, if you prefer, a multi-set of non-negative integers). A move is to choose a heap and remove any number of counters from that heap. So, a p-heap NIM game is a disjunctive sum of p one-heap games. If we can determine the canonical form of the one-heap game, then we hope, by learning how to add these games, to be able to solve the sum.

The value of the heap of size 0 is $\{ \mid \} = 0$, and the heap of size 1 is $\{0 \mid 0\} = *$. What about the heap of size 2? Well, it's of the form $\{0, * \mid 0, *\}$. Neither option for Left (or Right) is dominated or reversible, so this is a new value that we now dub $*2$ (pronounced "star two"). A heap of size 3 is $\{0, *, *2 \mid 0, *, *2\}$, and again there are no dominated or reversible options. This game is $*3$. In general, the values are called *nimbers* (or, sometimes more informally, *stars*).

Definition 7.3. The value of a NIM heap of size n, $n \geq 0$, is the *nimber* $*n$.

Note that $*0 = 0$ and $*1 = *$, and we will usually use the shorter names for these two games.

Theorem 7.4. *For $k > 0$, the canonical form of $*k$ is*

$$*k = \{0, *, *2, \ldots, *(k-1) \mid 0, *, *2, \ldots, *(k-1)\}.$$

Proof: See the next exercise.

Exercise 7.5. Prove Theorem 7.4 by showing that there are no dominated or reversible options. (*Hint:* First show that $*i$ and $*j$ are incomparable whenever $i \neq j$.)

We will now see that if the left and right options of a game G are equal and consist of just nimbers, then there is an easy way to find the canonical form and the value of G.

Definition 7.6. The *minimum excluded value* or *mex* of a set of non-negative integers is the least non-negative integer that does not occur in the set. This is denoted by $\text{mex}\{a, b, c, \ldots, k\}$.

For example, $\text{mex}\{0, 1, 2, 4, 5, 7\} = 3$, $\text{mex}\{1, 2, 4, 5, 7\} = 0$, and $\text{mex}\{\ \} = 0$.

Theorem 7.7. *Let* $G = \{*l_1, *l_2, \ldots, *l_k \mid *r_1, *r_2, \ldots, *r_j\}$ *and suppose that* $\text{mex}\{l_1, l_2, \ldots l_k\} = \text{mex}\{r_1, r_2, \ldots, r_j\} = n$. *Then,* $G = *n$ *and, consequently,* G *has the canonical form* $\{0, *, *2, \ldots, *(n-1) \mid 0, *, *2, \ldots, *(n-1)\}$.

Proof: We will show that $G - *n = 0$.

If either player moves either component to $*k$ for $k < n$, there is a matching move in the other component. In particular, since the mex of the $\{l_i\}$ and $\{r_i\}$ are both n, we have that $*k$ is both a left and a right option from G (and is also an option from $*n$). Hence, the second player can respond to $*k - *k = 0$.

The only other moves are from $G - *n$ to $*k - *n$ for $k > n$. In this case, $*n$ is an option from $*k$, so the second player responds locally to $*n - *n = 0$. □

Though Theorem 7.7 may appear mundane, its implications are truly profound:

Corollary 7.8. *Every impartial game is equivalent to a* NIM *heap. That is, for every impartial game* G *there is a non-negative integer* n *such that* $G = *n$.

Proof: Let G be impartial. By induction, the options of G are equivalent to NIM heaps. Theorem 7.7 then gives the recipe for finding the equivalent NIM heap for G. □

Exercise 7.9.

(a) Find k so that $G = \{0, *3, *4, *8, *2 \mid 0, *6, *4\} = *k$ using Theorem 7.7.

(b) Confirm that the theorem gave the correct k by playing the difference game $\{0, *3, *4, *8, *2 \mid 0, *6, *4\} - *k$.

Adding nimbers will be unusual. Already, we know that $* + * = 0$ and now from Exercise 7.1, in general, $*k + *k = 0$ for all $k \geq 0$. To find the *correct* way to add nimbers, we need to complete the analysis of NIM. We do this first with no mention of the values of the game or its summands.

7.2   The Analysis of Nim

Denote the game of NIM played with heaps of size a, b, ..., k by $\text{NIM}(a, b, \ldots, k)$. The *nim-sum* of numbers a, b, ..., k, written $a \oplus b \oplus \cdots \oplus k$, is obtained by adding the numbers in binary without carrying.[2] For example, here is the computation of the nim-sum of $12 \oplus 13 \oplus 7$:

$$
\begin{array}{r|cccc}
12 & 1 & 1 & 0 & 0 \\
13 & 1 & 1 & 0 & 1 \\
7 &   & 1 & 1 & 1 \\
\hline
   & 0 & 1 & 1 & 0
\end{array}
$$

and 0110 is binary for 6. For each column of bits (binary digits), if the column has an even number of 1s, the column sum is 0; if odd, the column sum is 1.

Theorem 7.10. *Let a, b, and c be non-negative integers. The nim-sum operation satisfies the following properties:*

- *commutativity: $a \oplus b = b \oplus a$;*

- *associativity: $(a \oplus b) \oplus c = a \oplus (b \oplus c)$;*

- *$a \oplus a = 0$; and*

- *$a \oplus b \oplus c = 0$ if and only if $a \oplus b = c$.*

Exercise 7.11. Prove Theorem 7.10.

We are now ready to reveal the secret behind winning NIM.

Theorem 7.12. *Let $G = \text{NIM}(a, b, \ldots, k)$. Then, G is a $\mathcal{P}$-position if and only if*

$$a \oplus b \oplus \cdots \oplus k = 0.$$

Proof: The proof of this result is an application of Theorem 2.15. That is, it suffices to show two things:

- If $a \oplus b \oplus \cdots \oplus k = 0$, then every move is to a position whose nim-sum is not zero.

- If $a \oplus b \oplus \cdots \oplus k \neq 0$, then some move exists to a position whose nim-sum is zero.

Suppose that $a \oplus b \oplus \cdots \oplus k = 0$. Without loss of generality, suppose that the first player removes r counters from heap a. Since the binary expansion of $a - r$ is not the same as a then $(a - r) \oplus b \oplus \cdots \oplus k \neq 0$.

---

[2]In computer science, the nim-sum operation is called the *exclusive-or*, or *xor* for short.

Now suppose that $q = a \oplus b \oplus \cdots \oplus k \neq 0$. Let $q_j q_{j-1} \ldots q_0$ be the binary expansion of q; that is, each bit q_i is either 1 or 0 and $q_j = 1$. Then, one of the heaps, again without loss of generality say a, must have a 1 in position j in its binary expansion.

We will show that there is a move from this heap of size a that produces a position whose nim-sum is 0; in particular, reducing a to $x = q \oplus a$.

Move is legal: In changing a to $q \oplus a$, the leftmost bit in a that is changed is a 1 (to a 0); therefore, $q \oplus a < a$, and so the move is legal. (The move *reduces* the size of the heap.)

Nim-sum 0: The resulting position is

$$(q \oplus a) \oplus b \oplus \cdots \oplus k = ((a \oplus b \oplus \cdots \oplus k) \oplus a) \oplus b \oplus \cdots \oplus k$$
$$= (a \oplus a) \oplus (b \oplus b) \oplus \cdots \oplus (k \oplus k)$$
$$= 0. \qquad \square$$

There are some observations that follow from this proof.

- If there is only a single heap left, the strategy of the theorem and common sense agree — the winning move is to take everything.

- If there are two heaps then the winning move, if there is one, is to make the two heaps the same size and afterward play Tweedledum-Tweedledee. The theorem gives the same strategy since $n \oplus n = 0$.

- The strategy applies even if the heap sizes are allowed to be transfinite. See Problem 7.

Example 7.13. Suppose that we are playing NIM(7, 12, 13):

12	1	1	0	0
13	1	1	0	1
7		1	1	1
6	0	1	1	0

Since the nim-sum is not 0, this is an $\mathcal{N}$-position, and the person on move can win. In particular, above the leftmost 1 in the sum, there is a 1 in *every* heap, and so there is a winning move from every heap. The player on move can reduce

- the heap of size 12 to one of size $12 \oplus 6 = 10$, or

- the heap of size 13 to one of size $13 \oplus 6 = 11$, or

- the heap of size 7 to one of size $7 \oplus 6 = 1$.

Exercise 7.14. Find all winning moves, if any, in the following.

(a) NIM$(3, 5, 7)$;

(b) NIM$(2, 3, 5, 7)$;

(c) NIM$(2, 4, 8, 32)$;

(d) NIM$(2, 4, 10, 12)$.

7.3 Adding Stars

All the hard work was done in the last section. We can now wrap everything up.

Theorem 7.15. *For non-negative integers k and j, $*k + *j = *(k \oplus j)$.*

Proof: From Theorem 7.12, we have that NIM with heaps of size k, j, and $k \oplus j$ is a $\mathcal{P}$-position. The values of the individual NIM heaps are, respectively, $*k$, $*j$, and $*(k \oplus j)$. That they form a $\mathcal{P}$-position means that

$$*k + *j + *(k \oplus j) = 0, \quad \text{or} \quad *k + *j = *(k \oplus j). \qquad \square$$

Corollary 7.16. $*a + *b + \cdots + *n = *(a \oplus b \oplus \cdots \oplus n)$.

Exercise 7.17. The game of POKER NIM is played with a collection of NIM heaps, with one additional bag with a finite number of counters. As a turn, a player can either make a normal NIM play, removing (and discarding) counters from one heap, *or* the player can increase the size of one heap by drawing counters from the bag.

Show that a winning strategy for POKER NIM is the same as in NIM. That is, if a player can win at NIM, the player can also win at POKER NIM by making the identical winning move(s).

Sometimes even a loopy, partizan game is really a finite, impartial game in disguise:

Exercise 7.18. Consider a second variant of POKER NIM from Exercise 7.17 in which each player has her own bag of counters. You can move either by removing counters from a single heap and placing them in your bag, or by taking counters from your bag and placing them on a single heap.

(a) As in Exercise 7.17, prove that a winning strategy for NIM suffices for playing this game.

(b) Explain why the game is *loopy*. Also explain why, nonetheless, a game played between talented players should end.

Example 7.19. SLIDING is played on a strip of squares numbered 1, 2, 3, . . ., n. There are several coins on the squares, at most one per square. A move is to slide a coin from its square to a lower-numbered square. The coin is not allowed to jump over or land on a square already occupied by a coin. Coins are not allowed to slide off the end of the strip.

For example, in the position above, the coin on square 7 can only move to 6, and the coin on 5 can move to squares 4, 3, or 2. The coin on 1 cannot move.

While this game, too, is equivalent to NIM, the heap sizes are not the positions of the coins. Rather, starting from the highest-numbered occupied square, pair up the coins. The NIM heap sizes are then the number of empty squares between each adjacent pair. If the number of coins is odd, then the leftmost coin at position i is tantamount to a heap of size $i - 1$.

For example, if the coins were on 15, 8, 7, 5, then pair up 15 with 8 and 7 with 5. We claim that this is equivalent to NIM with heaps of size 6 (gap between 15 and 8) and 1 (gap between 7 and 5), and the player on move can win by moving 15 to 10 (leaving gaps of 1 and 1). If the coins were on 15, 13, 12, 5, and 3, then we would pair 15-13, pair 12-5, and leave 3 unpaired; the heaps would be of sizes 1, 6, and 2. Since $*6 + * + *2 = *5$, there is at least one winning move. In NIM, one would replace $*6$ by $*3$, which is accomplished by moving the coin on square 12 to square 9.

Clearly, any heap can be decreased to any smaller size by moving the coin on the higher-numbered square without changing the size of any other heap. A heap may be increased in size if the coin on the lower square slides. However, there are only a finite number of moves that increase the size of a heap and, as in POKER NIM, these moves are of no consequence.

Exercise 7.20. Consider a variant of SLIDING in which the players are allowed to slide coins off the strip. Find the winning strategy of this game.

The nimbers have many other interesting properties. For one, they can be extended to a field (see *ONAG* [Con01] for example), but such a development is beyond the scope of this book. One feature, though, is that in the nimber $*k$, k does not have to be finite. Problem 7 shows such nimbers in action.

7.4 A More Succinct Notation

The original notation for impartial games was developed before the theory for partizan games was known. If only impartial games are under consideration, the

n	Options	Nim-values	mex	$\mathcal{G}(n)$
0	$\{\,\}$	$\text{mex}\{\,\}$		0
1	$\{0\}$	$\text{mex}\{\mathcal{G}(0)\}$	$\text{mex}\{0\}$	1
2	$\{1,0\}$	$\text{mex}\{\mathcal{G}(1),\mathcal{G}(0)\}$	$\text{mex}\{0,1\}$	2
3	$\{2,1\}$	$\text{mex}\{\mathcal{G}(2),\mathcal{G}(1)\}$	$\text{mex}\{1,2\}$	0
4	$\{3,2,0\}$	$\text{mex}\{\mathcal{G}(3),\mathcal{G}(2),\mathcal{G}(0)\}$	$\text{mex}\{0,2,0\}$	1
5	$\{4,3,1\}$	$\text{mex}\{\mathcal{G}(4),\mathcal{G}(3),\mathcal{G}(1)\}$	$\text{mex}\{1,0,1\}$	2
6	$\{5,4,2\}$	$\text{mex}\{\mathcal{G}(5),\mathcal{G}(4),\mathcal{G}(2)\}$	$\text{mex}\{2,1,2\}$	0

Table 7.1. Calculating the first few nim-values $\mathcal{G}(n)$ of a heap of size n appearing in SUBTRACTION$(1,2,4)$.

old notation is terser and more convenient. For the remainder of the chapter, we will therefore adopt it.

In this notation, since the left and right options are the same, they are not repeated. A nimber is abbreviated by writing k rather than $*k$. Specifically,

Definition 7.21. Let G be an impartial game. Then the *nim-value* (or *Grundy-value*), denoted by $\mathcal{G}(G)$, is k if and only if $G = *k$.

Observation 7.22. Let G be an impartial game and suppose that it has exactly n options. Then $\mathcal{G}(G) \leq n$.

Exercise 7.23. Prove this observation.

The term *Grundy-value* was the original term for $\mathcal{G}(G)$. More recently, it has been called the *nim-value* since if $G = *k$ then $\mathcal{G}(G) = k$ is the size of the NIM heap to which G is equivalent. We restate the central results for impartial games given in the last sections using this new notation:

Theorem 7.24.

- $\mathcal{G}(G) = \text{mex}\{\mathcal{G}(H) \mid H \text{ is an option of } G\}$.

- *If G, H, and J are impartial games, then $G = H + J$ if and only if $\mathcal{G}(G) = \mathcal{G}(H) \oplus \mathcal{G}(J)$.*

Example 7.25. The game SUBTRACTION$(1,2,4)$ is played with heaps of counters. A move consists of choosing a heap and removing either one, two, or four counters from that heap.

Table 7.1 shows how the first few nim-values of this game are calculated; a pattern seems apparent. Proving that such patterns hold is typically a rote induction argument.

Claim 7.26. *In SUBTRACTION$(1,2,4)$, the nim-value of a heap of size n is given by $\mathcal{G}(n) = n \pmod 3$.*

Proof: $\mathcal{G}(n) = \text{mex}\{\mathcal{G}(n-1), \mathcal{G}(n-2), \mathcal{G}(n-4)\}$

$$= \text{mex} \begin{cases} n-1 \pmod 3, \\ n-2 \pmod 3, \\ n-4 \pmod 3 \end{cases} \qquad \text{(by induction)}$$

$$= \text{mex} \begin{cases} n-1 \pmod 3, \\ n-2 \pmod 3 \end{cases} \qquad \text{(since } n-1 \equiv n-4 \pmod 3)$$

$$= n \pmod 3.$$

For this last step, the three possible cases are $\text{mex}\{0,1\} = 2$, $\text{mex}\{1,2\} = 0$, and $\text{mex}\{2,0\} = 1$. The base cases for $n \le 3$ are confirmed in Table 7.1. $\square$

Let G be the game with heaps of size 3, 4, 15, and 122. Now $\mathcal{G}(3) = 0$, $\mathcal{G}(4) = 1$, $\mathcal{G}(15) = 0$, and $\mathcal{G}(122) = 2$. Consequently, $\mathcal{G}(G) = 0 \oplus 1 \oplus 0 \oplus 2 = 3$, and so G is in $\mathcal{N}$. A winning move is one that reduces the nim-value of the game to 0, and surely such a move exists from 122, changing its nimber from 2 to 1. Remove one from the heap, leaving a heap of size 121.

In general, other winning moves might exist, similar to increasing a heap size in POKER NIM. Here, one might also change a nimber 1 to 2 or a nimber 0 to 3. The latter is impossible, for 3 does not appear as a nimber in the game. But, another winning move is to remove two from the heap of size 4 leaving 2.

Exercise 7.27. Find the nim-values for SUBTRACTION(1, 4).

7.5 Taking-and-Breaking Games

There are many natural variations on NIM obtained by modifying the legal moves. For example, sometimes a player, in addition to taking counters, might also be permitted to split the remaining heap into two (or sometimes more) heaps. These rule variants yield a rich collection of *Taking-and-Breaking games* that are discussed in *WW* [BCG01].

After NIM, Taking-and-Breaking games are among the earliest and most studied impartial games; however, by no means is everything known. To the contrary, much of the field remains wide open. Values in games such as GRUNDY'S GAME (choose a heap and split it into two different-sized heaps) and Conway's COUPLES ARE FOREVER (choose a heap and split it but heaps of size 2 are not allowed to be split) have been computed up to heaps of size 11×10^9 and 5×10^7, respectively, yet there is no complete analysis for these games.

In Taking-and-Breaking variants, the legal moves may vary with the size of the heap and the history of the game. For example, the legal moves might be, "a player must take at least one-quarter of the heap and no more than one-half," or "a player must take between $\frac{n}{2}$ and $n + 3$, where n is the number taken on the last move." For an example, see Problem 11. Games whose allowed moves depend on the history of the game are typically more difficult to analyze, but when the legal moves are independent of the history (and of moves in other heaps), then the game is a disjunctive sum and we only need to analyze games that have a single heap!

Definition 7.28. For a given Taking-and-Breaking game G, let $\mathcal{G}(n)$ be the nim-value of the game played with a heap of size n. The *nim-sequence* for the game is $\mathcal{G}(0)$, $\mathcal{G}(1)$, $\mathcal{G}(2)$, ...

In order to automate the process for finding (and proving) the nim-sequences of selected Taking-and-Breaking games, we need to address two main questions:

1. What types of regularities occur in nim-sequences?

2. When do we know that some regularity observed in a nim-sequence will repeat for eternity?

There are three types of regularities that have been observed in many nim-sequences to which we can answer the second question. These are listed in the next definition, but we only consider two of the three, those that are periodic and arithmetic periodic, in this book. The reader interested in split-periodicity should read [HN03, HN04].

Definition 7.29. A nim-sequence is

- *periodic* if there is some $l \geq 0$ and $p > 0$ such that $\mathcal{G}(n + p) = \mathcal{G}(n)$ for all $n \geq l$;

- *arithmetic periodic* if there is some $l \geq 0$, $p > 0$, and $s > 0$ such that $\mathcal{G}(n + p) = \mathcal{G}(n) + s$ for all $n \geq l$;[3] and

- *sapp regular* (or *split arithmetic periodic/periodic*) if there exist integers $l \geq 0$, $s > 0$, $p > 0$, and a set $S \subseteq \{0, 1, 2, \ldots, p - 1\}$ such that for all $n \geq l$,

$$\mathcal{G}(n + p) = \begin{cases} \mathcal{G}(n) & \text{if } n \ (\text{mod } p) \in S, \\ \mathcal{G}(n) + s & \text{if } n \ (\text{mod } p) \notin S. \end{cases}$$

The subsequence $\mathcal{G}(0), \mathcal{G}(1), \ldots, \mathcal{G}(l-1)$ is called the *pre-period* and its elements are the *exceptional values*. When l and p are chosen to be as small as possible, subject to meeting the conditions of the definition, we say that l is the *pre-period length* and p is the *period length*, while s is the *saltus*. If there is no pre-period,

---

[3]Reminder: + means normal, not nimber, addition!

the nim-sequence is called *purely periodic, purely arithmetic periodic,* or *purely sapp regular,* respectively.

Exercise 7.30. Match one-to-one each sequence on the left to a category on the right:

1231451671 ...	periodic
1123123123 ...	purely periodic
1122334455 ...	sapp regular
0123252729 ...	arithmetic periodic
0120120120 ...	purely sapp regular
1112233445 ...	purely arithmetic periodic

In each case, identify the period and (when non-zero) the saltus and pre-period.

7.6 Subtraction Games

In NIM we can remove any number of counters from a single heap. There are many natural variations where this freedom is somehow restricted. Perhaps we can only remove 2 or 3 counters from a heap, or perhaps we can remove any number of counters except 42 from a heap. In this section we will formally define and explore some games of this type — many of which we have seen in some examples and exercises:

Definition 7.31. A *subtraction game,* denoted SUBTRACTION(S), is played with heaps of counters and a set S of positive integers. A move is to choose a heap and remove any number of counters provided that number is in S.

- If $S = \{a_1, a_2, \ldots, a_k\}$ is finite, we have a *finite subtraction game,* which we denote SUBTRACTION($a_1, a_2, \ldots, a_k$).

- If, on the other hand, $S = \{1, 2, 3, \ldots\} \setminus \{a_1, a_2, \ldots, a_k\}$ consists of all the positive integers except a finite set, we have an *all-but subtraction game,* denoted ALLBUT($a_1, a_2, \ldots, a_k$).

In Example 7.25, SUBTRACTION(1, 2, 4) was shown to be periodic. On the other hand, ALLBUT() is another name for NIM and is arithmetic periodic with saltus 1.

Finite subtraction games

The following table gives the first 15 values of the nim-sequences for several subtraction games SUBTRACTION(S):

S	0	1	2	3	4	5	6	7	8	9	10	11	12	13	14
{1,2,3}	0	1	2	3	0	1	2	3	0	1	2	3	0	1	2
{2,3,4}	0	0	1	1	2	2	0	0	1	1	2	2	0	0	1
{3,4,5}	0	0	0	1	1	1	2	2	0	0	0	1	1	1	2
{3,4,6,10}	0	0	0	1	1	1	2	2	2	0	3	3	1	4	0

It is not surprising that the nim-sequence for SUBTRACTION$(1, 2, 3)$ is purely periodic with values $\dot{0}12\dot{3}$. (*Note:* We use dots to indicate the first and last values in the period.) The nim-sequence for SUBTRACTION$(2, 3, 4)$ is $\dot{0}0112\dot{2}$ and $\dot{0}00112\dot{2}$ for SUBTRACTION$(3, 4, 5)$. The pattern is not yet evident for SUBTRACTION$(3, 4, 6, 10)$, but if we pushed on we would eventually find that the nim-sequence is 000111222033140201312.

Calculating nim-sequences with a Grundy scale

The calculation of nim-sequences for subtraction games can be done by a computer (CGSuite, Maple, or Mathematica, for example, or any spreadsheet program), but there is an easy hand technique as well. Use two pieces of graph paper to construct what is known as a *Grundy scale*. Here, we will work with the example SUBTRACTION$(3, 4, 6, 10)$.

To begin, take two sheets of lined or graph paper. On each sheet make a scale, with the numbers marked in opposite directions. One of the scales is marked with △s to indicate the positions of numbers in the subtraction set; you can put a ▲ to indicate 0. The other sheet will be used to record the nim-values:

0	1	2	3	4	5	6	7

△				△		△	△			▲
11	10	9	8	7	6	5	4	3	2	1	0

As you fill in nim-values on the top scale, slide the bottom scale to the right. Shown below is the calculation $\mathcal{G}(9)$; take the mex of the positions marked by arrows. In this case, $\mathrm{mex}(1, 1, 2) = 0$, so we will fill in a 0 for $\mathcal{G}(9)$, now indicated by the ▲:

0	1	2	3	4	5	6	7	8	9	10	11	12	13	14
0	0	0	1	1	1	2	2	2	0					

△				△		△	△			▲
11	10	9	8	7	6	5	4	3	2	1	0

Similarly, $\mathcal{G}(13) = \mathrm{mex}(1, 2, 0, 3) = 4$:

0	1	2	3	4	5	6	7	8	9	10	11	12	13	14	15	16	17	18	19
0	0	0	1	1	1	2	2	2	0	3	3	1	4						

	△				△		△	△			▲
11	10	9	8	7	6	5	4	3	2	1	0

Exercise 7.32. Make a photocopy of the last Grundy scale, make a cut to separate the bottom and top portions, and use it to calculate more of the nim-sequence of SUBTRACTION$(3, 4, 6, 10)$. For more practice, try using Grundy scales for the examples in Problem 2 until you are comfortable.

Periodicity of finite subtraction games

After working out a few finite subtraction games, it will come as no surprise that their nim-sequences are always periodic.

Theorem 7.33. *The nim-sequences of finite subtraction games are periodic.*

Proof: Consider the finite subtraction game SUBTRACTION$(a_1, a_2, \ldots, a_k)$ and its nim-sequence. From any position there are at most k legal moves. So, using Observation 7.22, $\mathcal{G}(n) \leq k$ for all n.

Define $a = \max\{a_i\}$. Since $\mathcal{G}(n) \leq k$ for all n, there are only finitely many possible blocks of a consecutive values that can arise in the nim-sequence. So, we can find positive integers q and r with $a \leq q < r$ such that the a values in the nim-sequence immediately preceding q are the same as those immediately preceding r. Then, $\mathcal{G}(q) = \mathcal{G}(r)$ since

$$\mathcal{G}(q) = \mathrm{mex}\{\mathcal{G}(q - a_i) \mid 1 \leq i \leq k\} = \mathrm{mex}\{\mathcal{G}(r - a_i) \mid 1 \leq i \leq k\} = \mathcal{G}(r).$$

In fact, for such q and r and all $t \geq 0$, $\mathcal{G}(q+t) = \mathcal{G}(r+t)$. This is easily shown by induction — we have just seen the base case, and the inductive step is really just an instance of the base case translated t steps forward. Now set $l = q$ and $p = r - q$, and we see that the above says that, for all $n \geq l$, $\mathcal{G}(n + p) = \mathcal{G}(n)$; that is, the nim-sequence is periodic. $\qquad\square$

This proof shows that the pre-period and period lengths are at most $(k+1)^a$. However, this is generally a wild overestimate, and using the following corollary, the values of the period and pre-period lengths can usually be determined by computer:

Corollary 7.34. *Let $G = $ SUBTRACTION$(a_1, a_2, \ldots, a_k)$ and let $a = \max\{a_i\}$. If l and p are positive integers such that $\mathcal{G}(n) = \mathcal{G}(n+p)$ for $l \leq n < l+a$, then the nim-sequence for G is periodic with period length p and pre-period length l.*

Proof: See Problem 5.   □

That is, given conjectured values of l and p, it suffices to inspect the values of $\mathcal{G}(n)$ for $n \in \{l, l+1, \ldots, l+p+a-1\}$ to confirm the periodicity! It is then rote for a computer to identify the smallest pre-period and period given by the corollary.

Applying the corollary to the games in the previous table, we see that for

SUBTRACTION$(1,2,3)$ we have $l = 0$, $p = 4$, and $a = 3$, and these values can be confirmed by inspection of $\mathcal{G}(n)$ for $n \in \{0, \ldots, 6\}$;

SUBTRACTION$(2,3,4)$ $l = 0$, $p = 6$, and $a = 4$, check $n \in \{0, \ldots, 9\}$;

SUBTRACTION$(3,4,5)$ $l = 0$, $p = 8$, and $a = 5$, check $n \in \{0, \ldots, 12\}$; and

SUBTRACTION$(3,4,6,10)$ $l = 14$, $p = 7$, and $a = 10$, check $n \in \{14, \ldots, 30\}$.

Exercise 7.35. Use a Grundy scale to compute values of SUBTRACTION$(2,4,7)$ until you have enough to apply Corollary 7.34. For more practice, see Problem 2.

Two of the main questions, which still attract researchers, are:

1. How long can the period of SUBTRACTION$(a_1, a_2, \ldots, a_k)$ be as a function of the $\{a_i\}$?

2. Find general forms for the nim-sequence for SUBTRACTION(a_1, a_2, a_3).

Many games with different subtraction sets are actually the same in the sense that they have the same nim-sequence. For example, in SUBTRACTION(1), the first player wins precisely when there is an odd number of counters in the heap. The odd-sized heaps only have moves to even-sized heaps, even-sized heaps only have moves to odd-sized heaps, and the end position consists of heaps of size 0, that is, even. Therefore, a heap with n-counters is in $\mathcal{N}$ if n is odd, otherwise it is in $\mathcal{P}$. Adjoining any odd numbers to the subtraction set does not change this argument, and it is easy to show that the nim-sequence of any game SUBTRACTION$(1, \text{odds})$ is $\dot{0}\dot{1}0101$. A more general version of this analysis is sufficient to prove the following:

Theorem 7.36. *Let $G = $ SUBTRACTION$(a_1, a_2, \ldots, a_k)$ be purely periodic with period p. Let $H = $ SUBTRACTION$(a_1, a_2, \ldots, a_k, a_1 + mp)$ for $m \geq 0$. Then, G and H have the same nim-sequence.*

Proof: See Problem 6.   □

All-but-finite subtraction games

The following table gives the first 15 values of the nim-sequence for ALLBUT(S) for several sets S:

S	0	1	2	3	4	5	6	7	8	9	10	11	12	13	14
{1,2,3}	0	0	0	0	1	1	1	1	2	2	2	2	3	3	3
{2,3,4}	0	1	0	1	0	1	2	3	2	3	2	3	4	5	4
{1,2,8,9,10}	0	0	0	1	1	1	2	2	2	3	0	3	4	1	4

Grundy scales can still be used for the all-but subtraction games. This time, the small arrows mark the heaps that are not options.

Exercise 7.37. Use a Grundy scale to find the first 20 terms in the nim-sequence of ALLBUT$(1, 3, 4)$.

These values certainly do not look periodic, for $\mathcal{G}(n)$ appears to steadily increase with n, as confirmed by the following lemma:

Lemma 7.38. Let $G = $ ALLBUT$(a_1, a_2, \ldots, a_k)$ and $a = \max\{a_i\}$. Then, $\mathcal{G}(n + t) > \mathcal{G}(n)$ for all $t > a$.

Proof: Since any option from n is also an option from $n + t$, $\mathcal{G}(n + t) \geq \mathcal{G}(n)$. Additionally, n is an option from $n + t$, so $\mathcal{G}(n)$ occurs as the nim-value of an option from $n + t$ and thus we do not have equality. $\square$

However, as we will see, all-but subtraction games are arithmetic periodic. From the previous table, we might conjecture the following (we denote the saltus information in parentheses with a + sign):

S	Nim-sequence
{1,2,3}	$\dot{0}00\dot{0}(+1)$
{2,3,4}	$\dot{0}1010\dot{1}(+2)$
{1,2,8,9,10}	$00011122\dot{2}3\dot{0}(+1)$

Proving that a nim-sequence is arithmetic periodic is typically more difficult than proving periodicity. Nonetheless, we can parallel the work done for finite subtraction games, proving first that ALLBUT subtraction games are arithmetic periodic and then identifying how one (a person or a computer) can automatically confirm the arithmetic periodicity.

Theorem 7.39. Let $G = $ ALLBUT$(a_1, a_2, \ldots, a_k)$. Then, the nim-sequence for G is arithmetic periodic.

Proof: Lemmas 7.41 and 7.43, motivated and then proved below, yield the theorem. $\square$

The proof of this theorem is more technical than that of Theorem 7.33, but in broad strokes it is similar. In the proof of Theorem 7.33, we first argued that nim-values cannot get too large and therefore that some sufficiently long sequence of nimbers must repeat. Once a sufficiently long sequence appears identically twice, an induction argument establishes that those two sequences remain in lock-step ad infinitum.

With arithmetic periodicity, the repetition that we seek is in the *shape* of a sequence rather than its values, as shown in Figure 7.1 on page 172. As you read through the proofs, keep in mind that two subsequences of nim-values, call them $(\mathcal{G}(n_0), \dots, \mathcal{G}(n_0 + c))$ and $(G(n_0'), \dots, \mathcal{G}(n_0' + c))$, have the same *shape* if

1. the two subsequences differ by a constant,

$$\mathcal{G}(n_0') - \mathcal{G}(n_0) = \mathcal{G}(n_0' + 1) - \mathcal{G}(n_0 + 1) = \cdots = \mathcal{G}(n_0' + c) - \mathcal{G}(n_0 + c),$$

2. or equivalently, both subsequences move up and down the same way: for all $0 \leq i < c$,

$$\mathcal{G}(n_0 + i + 1) - \mathcal{G}(n_0 + i) = \mathcal{G}(n_0' + i + 1) - \mathcal{G}(n_0' + i).$$

It will turn out that the base case for our inductive proof will require a repetition of length about $2a$, where $a = \max\{a_i\}$. We show that such a repetition exists in the next two lemmas.

Lemma 7.40. *Let* $G = \text{ALLBUT}(a_1, a_2, \dots, a_k)$, *and define* $a = \max\{a_i\}$. *For all* $n \geq a$,

$$k - a \leq \mathcal{G}(n + 1) - \mathcal{G}(n) \leq a - k + 1.$$

Proof: Fix $n > a$ and let $X \subseteq \{\mathcal{G}(0), \mathcal{G}(1), \mathcal{G}(2), \dots, \mathcal{G}(n - 1)\}$ be the nim-values of the options of n.[4] Now, since $\mathcal{G}(n)$ is the mex of X, we know that $\{0, 1, 2, \dots, \mathcal{G}(n) - 1\} \subseteq X$. Further, play in $G = \text{ALLBUT}(a_1, a_2, \dots, a_k)$ prohibits moves to k of the top a heap sizes. Hence, one of $\{\mathcal{G}(0), \dots, \mathcal{G}(n - a - 1)\}$, say $\mathcal{G}(m)$, must be at least $\mathcal{G}(n) - 1 - (a - k)$, for only $a - k$ of the terms from X can appear among $\{\mathcal{G}(n - a), \dots, \mathcal{G}(n - 1)\}$. Also, m and all the options from m are also options from $n + 1$. Thus, we have that

$$\mathcal{G}(n + 1) > \mathcal{G}(m), \quad \text{and so}$$
$$\mathcal{G}(n + 1) \geq \mathcal{G}(n) - (a - k).$$

Similarly, for the second inequality, one of $\{\mathcal{G}(0), \dots, \mathcal{G}(n - a - 1)\}$ is at least $\mathcal{G}(n + 1) - 2 - (a - k)$, and it and its options are options of n. So,

$$\mathcal{G}(n) \geq \mathcal{G}(n + 1) - (a - k) - 1. \qquad \square$$

---

[4] In other words, $X = \{\mathcal{G}(n - \alpha) \mid \alpha \notin \{a_1, \dots, a_k\}\}$.

Lemma 7.41. *Let $G = \text{ALLBUT}(a_1, a_2, \ldots, a_k)$ and $a = \max\{a_i\}$. There exist n_0, n_0', s, and $p = n_0' - n_0 > 0$ such that $\mathcal{G}(n + p) - \mathcal{G}(n) = s$ for $n_0 \leq n \leq n_0 + 2a$.*

Proof: By Lemma 7.40, for all n, $\mathcal{G}(n + 1) - \mathcal{G}(n)$ must lie between $k - a$ and $a - k + 1$. But, there are only $2(a - k) + 2$ values in that range. Hence, setting $c = 2(a - k) + 2$, there are at most c^{2a} possible sequences of the form

$$\begin{aligned}
(\quad & \mathcal{G}(n + 1) - \mathcal{G}(n), \\
& \mathcal{G}(n + 2) - \mathcal{G}(n + 1), \\
& \ldots, \\
& \mathcal{G}(n + 2a) - \mathcal{G}(n + 2a - 1) \;),
\end{aligned} \tag{7.1}$$

and so, eventually, for two values $n = n_0$ and $n = n_0'$, the two corresponding sequences are identical. The lemma follows. □

Exercise 7.42. Complete the algebra to confirm that "the lemma follows." In particular, given the two identical sequences satisfying (7.1), one with n_0' and one with n_0, you need to define p and explain why $\mathcal{G}(n+p) - \mathcal{G}(n)$ is a constant (call it $s = \mathcal{G}(n_0') - \mathcal{G}(n_0)$) for $n_0 \leq n \leq n_0 + 2a$. (The matching shapes in Figure 7.1 provide some intuition.)

The next lemma completes the inductive step of the proof of Theorem 7.39, showing that once two sufficiently long sequences have the same shape, they are fated to continue in lock-step.

Lemma 7.43. *Let $G = \text{ALLBUT}(a_1, a_2, \ldots, a_k)$ and $a = \max\{a_i\}$, and suppose that $\mathcal{G}(n + p) - \mathcal{G}(n) = s$ for $n_0 \leq n \leq n_0 + 2a$. Then, $\mathcal{G}(n + p) - \mathcal{G}(n) = s$ for all $n \geq n_0$.*

In particular, $\mathcal{G}(n)$ has pre-period length $l = n_0$, period p, and saltus s, and one can identify and confirm the period and saltus by only inspecting the first $l + 2a + p + 1$ values; i.e., $\mathcal{G}(n)$ for $0 \leq n \leq l + 2a + p$.

Proof: Using induction, it suffices to prove that $\mathcal{G}(n + p) - \mathcal{G}(n) = s$ for $n = n_0 + 2a + 1$. Define the following quantities as shown in Figure 7.1: $n_0' = n_0 + p$, $n_1 = n_0 + a$, $n_2 = n_0 + 2a$, $n_1' = n_0' + a$, and $n_2' = n_0' + 2a$.

As we compute $\mathcal{G}(n') = \mathcal{G}(n+p)$ as the mex of the nim-values of its options, by Lemma 7.38 $\mathcal{G}(n')$ exceeds all $\mathcal{G}(m)$ for $m < n' - a - 1 = n_1'$. And we know that $\mathcal{G}(n') > \mathcal{G}(n_1')$ in any event, so we can safely ignore $\mathcal{G}(m)$ for $m < n_0'$. In other words, $\mathcal{G}(n')$ is the minimum excluded value from $\{\mathcal{G}(n_0'), \ldots, \mathcal{G}(n_2')\}$ that exceeds $\mathcal{G}(n_1')$. Since the assignment of $\mathcal{G}(n')$ is unaffected by linear translation of those nim-values, $\mathcal{G}(n') - \mathcal{G}(n) = s$. □

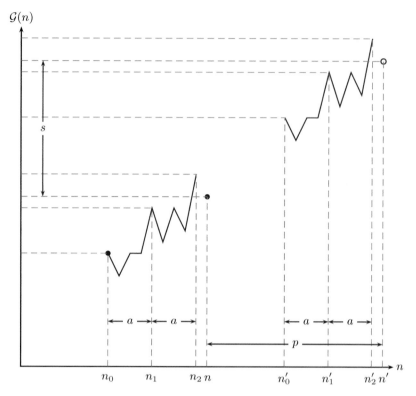

Figure 7.1. A diagram sketching the proof of arithmetic periodicity in ALLBUT subtraction games. For the game ALLBUT$(a_1, \ldots, a_k)$, $a = \max\{a_i\}$, s is the proposed saltus and p is the proposed period. We first find two sufficiently long sequences of nim-values with the same shape (but translated upward) and inductively prove that those sequences remain in lock-step.

   This last lemma gives an automated method for testing when an all-but subtraction game nim-sequence has become arithmetic periodic. Although Figure 7.1 shows the two subsequences non-overlapping (suggesting that $p > 2a$), the proof is unaffected by overlap.

Exercise 7.44. Apply Lemma 7.43 to find the period length p and the saltus s of the game from Exercise 7.37. In particular, how many values of $\mathcal{G}(n)$ need computing to confirm the period and saltus? (*Hint:* The game is purely arithmetic periodic with period between 10 and 15.)

Exercise 7.45. We asserted at the start of this section that the nim-sequence for ALLBUT$(1, 2, 8, 9, 10)$ is given by $00011122\dot{2}3\dot{0}(+1)$. To apply Lemma 7.43, which values of $\mathcal{G}(n)$ need be confirmed to be confident of the nim-sequence?

As was seen in the table on page 169, some ALLBUT subtraction games have pre-periods. If you do Problems 12, 13, and 14, then you will have shown that the nim-values of ALLBUT games where s has cardinality 1 and 2 are purely arithmetic periodic.

Frequently, the ALLBUT subtraction set can be reduced. While most such reductions remain specific to individual games, we do have one general reduction theorem.

Theorem 7.46. *Let $a_1 < a_2 < \cdots < a_k$ be positive integers, and let $b > 2a_k$. The nim-sequences of* ALLBUT$(a_1, a_2, \ldots, a_k)$ *and* ALLBUT$(a_1, a_2, \ldots, a_k, b)$ *are equal.*

Proof: Let $\mathcal{G}(n)$ denote the nim-sequence of ALLBUT$(a_1, a_2, \ldots, a_k)$ and $\mathcal{G}'(n)$ that of ALLBUT$(a_1, a_2, \ldots, a_k, b)$. Certainly, $\mathcal{G}(n) = \mathcal{G}'(n)$ for $n < b$ since the options in the two games are identical to that point. Suppose inductively that the two nim-sequences agree through $n - 1$, and consider $\mathcal{G}(n)$ and $\mathcal{G}'(n)$. Since the options of ALLBUT$(a_1, a_2, \ldots, a_k, b)$ are a subset of those of ALLBUT$(a_1, a_2, \ldots, a_k)$, the only possible way to have $\mathcal{G}(n) \neq \mathcal{G}'(n)$ would be if $\mathcal{G}'(n) = \mathcal{G}(n - b)$, since the latter is the only possible value that does not occur among the options of ALLBUT$(a_1, a_2, \ldots, a_k, b)$ but does occur among the options of ALLBUT$(a_1, a_2, \ldots, a_k)$. In order for this to be true, it would also be the case that no option of n in ALLBUT$(a_1, a_2, \ldots, a_k, b)$ had value $\mathcal{G}(n - b)$.

However, consider $m = n - b + a_k$. By the inductive hypothesis $\mathcal{G}(m) = \mathcal{G}'(m)$. Moreover, all values smaller than $n - b$ are options from a heap of size m in ALLBUT$(a_1, a_2, \ldots, a_k)$, so $\mathcal{G}(m) \geq \mathcal{G}(n - b)$. As m is an option of n in ALLBUT$(a_1, a_2, \ldots, a_k, b)$, in order to avoid a contradiction we would require that $\mathcal{G}(m) > \mathcal{G}(n - b)$. But then, m would have an option m' in ALLBUT$(a_1, a_2, \ldots, a_k)$ with $\mathcal{G}(m') = \mathcal{G}(n - b)$. Since $m' \neq n - b$ and $m' < n - a_k$, m' is also an option of n in ALLBUT$(a_1, a_2, \ldots, a_k, b)$. This contradiction establishes the desired result. $\qquad \square$

For more on all-but-finite subtraction games, see [Sie06].

Kayles and kin

KAYLES is played with a row of pins standing in a row. The players throw balls at the pins. The balls are only wide enough to knock down one or two adjacent pins.

Above is a game in progress. Would you like to take over for the next player?

KAYLES can also be thought of as a game played with heaps of counters and a player is allowed to take one or two counters and possibly to split the remaining heap into two. KAYLES is clearly a Taking-and-Breaking game, but maybe it is too hard right now. We will work up to it with a couple of simpler variants.

First, suppose that we play a game where a player is only allowed to split the heap into two non-empty heaps (no taking). We saw in Example 1.8 that this is SHE LOVES ME SHE LOVES ME NOT disguised. That is,

$$\mathcal{G}(n) = \begin{cases} 0 & \text{if } n = 0, \\ 0 & \text{if } n > 0 \text{ is odd,} \\ 1 & \text{if } n > 0 \text{ is even.} \end{cases}$$

In particular, moves from an even heap are to even + even $= * + * = 0$ or odd + odd $= 0$, while moves from an odd heap are to even + odd $= * + 0 = *$.

The version where you are allowed to split *or* take-one-and-must-split (into non-empty heaps) needs a bit more care. For a heap of n, the splitting move gives the options $n - i$ and i for $i = 1, 2, \ldots, n - 1$, and the take-one-and-split move leaves the positions $n - 1 - j$ and j, $j = 1, 2, \ldots, n - 2$. To find the value of the game, we need to find the mex of

$$\{\mathcal{G}(i) \oplus \mathcal{G}(n - i) \mid i = 1, 2, \ldots, n - 1\}$$
$$\cup \{\mathcal{G}(j) \oplus \mathcal{G}(n - 1 - j) \mid j = 1, 2, \ldots, n - 2\}.$$

The first few values are $\mathcal{G}(0) = 0$, $\mathcal{G}(1) = 0$, $\mathcal{G}(2) = \text{mex}\{\mathcal{G}(1) + \mathcal{G}(1)\} = 1$, $\mathcal{G}(3) = \text{mex}\{\mathcal{G}(2) \oplus \mathcal{G}(1)\} \cup \{\mathcal{G}(1) \oplus \mathcal{G}(1)\} = 2$. This becomes tedious very quickly, and a machine will be very useful. If one is not available, then a Grundy scale can help. This time both pieces of paper have the nim-values, one from left to right as before and the other from right to left. However, we have to use them twice and record the intermediate values. The next diagram shows the calculation of $\mathcal{G}(9)$. First, line up the scales for the split move, $8 + 1$, $7 + 2$, ..., $1 + 8$. Clearly we only need go to $4 + 5$ since we repeat the calculations in reverse. Nim-sum the pairs of numbers and record:

0	1	2	3	4	5	6	7	8	9	10	11	12	13	14
0	0	1	2	3	1	4	3	2						

			2	3	4	1	3	2	1	0	0
11	10	9	8	7	6	5	4	3	2	1	0

This gives $\{2, 2, 6, 2, 2, 6, 2, 2\}$ as the set of values for these options. Now line them up for the take-one-and-split move:

0	1	2	3	4	5	6	7	8	9	10	11	12	13	14
0	0	1	2	3	1	4	3	2						

			2	3	4	1	3	2	1	0	0
11	10	9	8	7	6	5	4	3	2	1	0

This gives the set $\{3, 5, 3, 0, 3, 5, 3\}$ for these options. The least non-negative number that does not appear in either set is 1, so record $\mathcal{G}(9) = 1$ on both papers:

0	1	2	3	4	5	6	7	8	9	10	11	12	13	14
0	0	1	2	3	1	4	3	2	**1**					

	1	2	3	4	1	3	2	1	0	0	
11	10	9	8	7	6	5	4	3	2	1	0

Exercise 7.47. Continue using the Grundy scale to compute five more nim-values of this game (where you are allowed to split *or* split-and-take-one but always-leave-two-heaps).

For a fixed (finite) set S of positive integers, we can define the variant SPLITTLES(S) where a player is allowed to take away s, for some $s \in S$, from a heap of size at least s and possibly split the remaining heap.

In particular, KAYLES is just SPLITTLES$(1, 2)$.

What sort of regularities could the nim-sequences of SPLITTLES(S) have? In the few examples so far, the nim-values do not grow very quickly at all, suggesting that the nim-sequences are periodic *but Nobody Knows*. All we know is that the nim-sequences are *not* arithmetic periodic [Aus76, BCG01]. It is believed that SPLITTLES(S) is periodic when S is finite, and for periodicity we do have an automatic check.

Theorem 7.48. *Fix* SPLITTLES(S) *with* $m = \max S$. *If there exist integers* $l \geq 0$ *and* $p > 0$ *such that* $\mathcal{G}(n + p) = \mathcal{G}(n)$ *for* $l \leq n \leq 2l + 2p + s$, *then* $\mathcal{G}(n + p) = \mathcal{G}(n)$ *for all* $n \geq l$. *That is, the period persists forever.*

Proof: See Problem 15. $\square$

7.7   Keypad Games

It is easy to modify a game so that playing a move blocks other moves. For example, we could play a subtraction game with the condition that the next player cannot repeat the last move. Consider this restriction when the subtraction set is $S = \{1, 2, 3\}$. In this case, we cannot tell what is legal by looking at the pile size, for we also need to know the previous play to know what is legal. So, we define a position that includes the information about the state of the game. Let $N(n) = (N(n, 1), N(n, 2), N(n, 3), N(n, 0))$ be the position where n is the heap size and the second number is the last move. The last entry in the tuple, $N(n, 0)$, represents the position if it is the first move of the game. The

	Last move	Heap 0	1	2	3	4	5	6	7	8	9
Outcomes	1	$\mathcal{P}$	$\mathcal{P}$	$\mathcal{N}$	$\mathcal{N}$	$\mathcal{P}$	$\mathcal{P}$	$\mathcal{N}$	$\mathcal{N}$	$\mathcal{P}$	$\mathcal{N}$
	2	$\mathcal{P}$	$\mathcal{N}$	$\mathcal{N}$	$\mathcal{N}$	$\mathcal{P}$	$\mathcal{N}$	$\mathcal{N}$	$\mathcal{N}$	$\mathcal{P}$	$\mathcal{N}$
	3	$\mathcal{P}$	$\mathcal{N}$	$\mathcal{N}$	$\mathcal{P}$	$\mathcal{P}$	$\mathcal{N}$	$\mathcal{N}$	$\mathcal{N}$	$\mathcal{P}$	$\mathcal{P}$
	0	$\mathcal{P}$	$\mathcal{N}$	$\mathcal{N}$	$\mathcal{N}$	$\mathcal{P}$	$\mathcal{N}$	$\mathcal{N}$	$\mathcal{N}$	$\mathcal{P}$	$\mathcal{N}$
Nim-values	1	0	0	1	2	0	0	1	1	0	0
	2	0	1	1	2	0	2	1	2	0	2
	3	0	1	1	0	0	1	1	0	0	1
	0	0	1	1	2	0	3	1	3	0	3

Table 7.2. Outcome classes and nim-values for the KEYPAD game $S = \{1, 2, 3\}$ and $T_i = S \setminus \{i\}$.

options are given by

$$N(n, 1) = \{N(n - 2, 2), N(n - 3, 3)\},$$
$$N(n, 2) = \{N(n - 1, 1), N(n - 3, 3)\},$$
$$N(n, 3) = \{N(n - 1, 1), N(n - 2, 2)\},$$
$$N(n, 0) = \{N(n - 1, 1), N(n - 2, 2), N(n - 3, 3)\}.$$

We computed outcomes and nim-values in Table 7.2. It looks like the outcomes are periodic from heap size 0 and the nim-values from heap size 2. Can we prove this?

A *keypad game* is a subtraction game with finite subtraction set $S = \{s_1, s_2, \ldots, s_k\}$. For every s_i there is an associated *move set* $T_i \subset S$. If subtracting s_i was the last move, then the next subtraction must be some $s \in T_i$.

The first game of this section is $S = \{1, 2, 3\}$ and $T_i = S \setminus \{i\}$. Richard Guy introduced *the* KEYPAD game where $S = \{0, 1, 2, 3, 4, 5, 6, 7, 8, 9\}$ and T_i consists of the numbers on a phone keypad

$$\begin{array}{|c|c|c|}
\hline
1 & 2 & 3 \\
\hline
4 & 5 & 6 \\
\hline
7 & 8 & 9 \\
\hline
\multicolumn{1}{c}{} & 0 & \multicolumn{1}{c}{} \\
\cline{2-2}
\end{array}$$

that are in the same row or column as i excluding i. For example, $T_0 = \{2, 5, 8\}$ and $T_1 = \{2, 3, 4, 7\}$. Yu and Banerji [YB82] introduced the DIE game that is played with a heap and a six-sided die; the next player rotates the die a quarter turn and subtracts the number on the top face from the heap. In this game, $S = \{1, 2, 3, 4, 5, 6\}$ and $T_i = S \setminus \{7 - i, i\}$, since opposite faces on the die sum to 7. At the start of the KEYPAD game, there is no previous move and all subtractions are legal. At the start of the DIE game, the die will be in some position.

Let K be a keypad game with subtraction set S and move sets $T_1, T_2, \ldots, T_k$. For the heap of size n, let $A_i(n)$ be the options where s_i was the previous move. Let $A(n) = (A_1(n), A_2(n), \ldots, A_k(n))$, then

$$A_{\mathcal{G}}(n) = (\mathcal{G}(A_1(n)), \mathcal{G}(A_2(n)), \ldots, \mathcal{G}(A_k(n)))$$

is the k-tuple of nim-values and

$$A_o(n) = (o(A_1(n)), o(A_2(n)), \ldots, o(A_k(n)))$$

is the k-tuple of outcomes. Let $A_{\mathcal{G}}(0), A_{\mathcal{G}}(1), \ldots$ be the *keypad value sequence* of nim-values and $A_o(0), A_o(1), \ldots$ be the *keypad outcome sequence* of outcomes.

Theorem 7.49. *The keypad sequences of keypad games are periodic.*

The proof is very similar to that of Theorem 7.33. We leave it to the reader to fill in the details. Also, the periodicity check, Corollary 7.34, can be modified for keypad games.

Corollary 7.50. *Let K be a keypad game with subtraction set $(s_1, s_2, \ldots, s_k)$, let $s = \max\{s_i\}$, and let the move sets of K be T_i. If l and p are positive integers such that $A_{\mathcal{G}}(n) = A_{\mathcal{G}}(n + p)$ for $l \leq n < l + s$, then the keypad value sequence for K is periodic with period length p and pre-period length at most l.*

The reader should also fill in the details of the proof of Corollary 7.50. Note that the Corollary can be modified so that "value" is replaced by "outcome."

Using Corollary 7.50, the information in Table 7.2 shows that, for our original game, the keypad outcome sequence is periodic with $l = 0$ and $p = 4$ and the keypad values are periodic with $l = 4$ and $p = 4$. With a bit of work, it is straightforward to show the following:

	Pre-period length	Period length
DIE	27	9
KEYPAD without 0	129	1
KEYPAD with 0	117	15

Exercise 7.51. If we wrote a computer program to compute KEYPAD nim-values, up to how large a pile size would we need to run the program to provably confirm the pre-periods and periods in the previous table?

Problems

1. Find the nim-sequences for SUBTRACTION(S), where $|S| = 2$ and $S \subseteq \{1, 2, 3, 4\}$.

2. Use a Grundy scale and Corollary 7.34 to compute the nim-sequences of

   (a) SUBTRACTION(2, 3, 5);

   (b) SUBTRACTION(3, 5, 8);

   (c) SUBTRACTION(1, 3, 4, 7, 8).

3. Find the period for

   (a) ALLBUT(1, 2, 3);

   (b) ALLBUT(5, 6, 7);

   (c) ALLBUT(3, 4, 6, 10).

4. A game is played like KAYLES, only you cannot bowl the end of a row of pins. In NIM language, you can take one or two counters from a heap and you must split that heap into two *non-empty* heaps. Using a Grundy scale, calculate the first 15 nim-values for this game.

5. Prove Corollary 7.34 on page 167.

6. Prove Theorem 7.36 on page 168.

7. (This is a generalization of NIM.) POLYNIM is played on polynomials with non-negative coefficients. A move is to choose a single polynomial and reduce one coefficient and arbitrarily change or leave alone the coefficients on the smaller powers in this polynomial — $3x^2 + 15x + 3$ can be reduced to $0x^2 + 19156x + 2345678 = 19156x + 2345678$. Analyze POLYNIM. In particular, identify a strategy for determining when a position is a $\mathcal{P}$-position analogous to Theorem 7.12.

8. Find the nim-sequences for SUBTRACTION$(1, 2q)$ for $q = 1, 2, 3$. Find the form of the nim-sequence of SUBTRACTION$(1, 2q)$ for arbitrary q.

9. Show that the nim-sequence for SUBTRACTION$(q, q+1, q+2)$ is $\dot{0}0^{q-1}1^q2\dot{2}$ if $q > 1$. (As usual, x^b is x repeated b times.)

10. Analyze SUBTRACTION$(1, 2, 4, 8, 16, \ldots, 2^i, \ldots)$.

11. Analyze this variant of NIM: on any move a player must remove at least half the number of counters from the heap.[5]

---

[5]The nim-sequence for the game in which no more than half can be removed has a remarkable self-similarity property: if you remove the first occurrence of each number in the nim-sequence, then the resulting sequence is the same as the original! See [Lev06].

12. Find the periods for ALLBUT(1), ALLBUT(2), and ALLBUT(3). Conjecture and prove your conjecture for the period of ALLBUT(q).

13. Find the periods for ALLBUT(1, 2), ALLBUT(2, 3), and ALLBUT(3, 4). Conjecture and prove your conjecture for the period of ALLBUT($q, q + 1$).

14. Find the nim-sequence for ALLBUT(q, r), $q < r$. (*Hint:* There are two cases $r = 2q$ and $r \neq 2q$.)

15. Prove Theorem 7.48 on page 175.

16. The rules of the game TURN-A-BLOCK are at the textbook website, www. lessonsinplay.com, and you can play the game against the computer. Determine a winning strategy for TURN-A-BLOCK. You should be able to consistently beat the computer on the *hard* setting at 3×3 and 5×3 gnturn-a-block (and even bigger boards!). You should be able to determine who should win from any position up to 5×5.

17. NO REPEAT NIM is the same as NIM except that the next turn cannot reduce the heap reduced on the previous turn.

    (a) Find the $\mathcal{P}$-positions for positions with one or two heaps.

    (b) Find all $\mathcal{P}$-positions.

Preparation for Chapter 8

Prep Problem 8.1. Redo Exercise 5.33 on page 117 to remind your-self of the definition of left and right stops.

To the instructor: We follow the non-constructive proof of the Mean Value Theorem of games, Theorem 8.6 on page 186. Our proof mirrors that of *ONAG* [Con01]. As an alternative, you may wish to present the constructive proof that appears in [BCG01, pp. 152–155]. Their approach uses thermographs, so you should present the proof after covering material from Section 8.3 on drawing thermographs.

Chapter 8

Hot Games

In many classical games, the issue of whose turn it is to move is critical either
in terms of determining the outcome or in terms of its influence on the final
score (assuming error-free play). The terminology of these games reflects the
importance of this concept: one of the fundamental strategic concepts in CHESS
is *tempo,* while GO distinguishes moves that keep *sente* (i.e., demand a response
and so preserve the initiative) from those that are *gote* (valuable, perhaps, but
not requiring a local reply).

In a limited sense, these concepts relate only to the outcome class of the
game in question. Specifically, if a game is of type $\mathcal{N}$ then both players would
like to have the opportunity to move next. However, there is certainly more to
it than that — players recognize these situations as being important exactly
when the value of the next move is such that it establishes a decisive advantage,
usually in terms of material gain in CHESS or gain of territory or a successful
capture in GO. The positions that are important in these analyses are more
like switches, such as $\pm 5 = \{5 \mid -5\}$, than stars.

Thermography is a technique that allows an understanding of some of the
complex issues that arise in playing sums of hot games, or games with several
hot options. Thermography is useful in the exact analysis of endgame positions in GO. Thermostrat and Sentestrat, modifications of the greedy strategy
based on thermography, show promise as tools in building effective computer
programs for playing combinatorial games.

8.1   Comparing Games and Numbers

The way in which a game G compares to an arbitrary number provides a certain amount of information about it — sometimes enough to establish the winner in more complicated positions. For example, if we know that $G \geq 1$, then Left will be happy about playing the game $H = G - 1 + \uparrow$ since she can see that H is positive. Roughly speaking, Left is interested in the answer to the question, "For which numbers x is $G \geq x$?" while Right would like to know, "For which numbers y is $G \leq y$?"

Theorem 8.1. *Let G be a short game. There exists a positive integer n such that*

$$-n < G < n.$$

Proof: This is an immediate corollary to Theorem 6.3, which states that the greatest and least games born by day n are n and $-n$, respectively. Simply take n to be one more than the birthday of G.   □

If Left and Right are playing a single short game G, then neither will be particularly keen to carry on when a position is reached whose value is a number. At this point it is clear who has won based simply on the sign of the number and, if the number is 0, whose turn it is. If we think of this number as the score of the game (a concept that we have been avoiding until now!), then Left will want to play to maximize the score and Right to minimize it. Note also that when the game reaches its stopping value, Left would rather that it were Right's turn to move (since a move by Right would then further improve the position for Left) and vice versa. This notion of score (or rather scores, since it depends on the initial player) has already been seen in Section 5.3 under the guise of the left and right stops of a game. Since this concept is central to our development here, we will recapitulate and extend it a bit, as well as consider some examples.

Consider the DOMINEERING position

$$G = \ \text{}$$

Either of Left's moves leave the value $*$. Right can choose to leave $*$ or -1 and clearly prefers the latter. So, $G = \{* \mid -1\}$. If Right plays first the stopping value of G is -1, and when G stops it will be Left's turn to move. So, the *right stop* of G is -1_R. Notice that we have adorned the *stopping value* -1 with a subscript to indicate who moved last. Similarly, if Left moves first the stopping value will be 0 (after Right plays in the $*$), and again it will be Left's turn to move. So, the *left stop* of G is 0_R.

Now consider a game $H = \{0, * \mid -1\}$. The right stop of H is clearly -1_R. What about the left stop? Well, regardless of which move Left chooses, the stopping value will be 0. However, if she moves to $*$ it will be her turn to play when the stop is reached, while if she moves to 0 it will be Right's turn. As we noted above, she prefers the latter alternative, so the left stop of H is 0_L.

Definition 8.2. An *adorned stop* (or just *stop*) at x is a symbol of the form x_R or x_L, where x is a number. Stops are totally ordered, using the underlying natural order for numbers and $\square_L > \square_R$ used for tie-breaks; i.e., if $x > y$, then

$$x_L > x_R > y_L > y_R.$$

Context will often determine whether we mean the adorned or unadorned stop, as the word *stop* and the notation $\mathbf{RS}(G)$ are ambiguous. If we want to emphasize the unadorned stop, we can just write the *value* of the stop x_L (or of x_R) as x.

We repeat the definition of left and right stops from page 117, which now can include adornment:

Definition 8.3. Denote the *left stop* and *right stop* of a game G by $\mathbf{LS}(G)$ and $\mathbf{RS}(G)$, respectively. They are defined in a mutually recursive fashion:

$$\mathbf{LS}(G) = \begin{cases} x_R & \text{if } G \text{ is the number } x, \\ \max(\mathbf{RS}(G^L)) & \text{if } G \text{ is not a number}; \end{cases} \tag{8.1}$$

$$\mathbf{RS}(G) = \begin{cases} x_L & \text{if } G \text{ is the number } x, \\ \min(\mathbf{LS}(G^R)) & \text{if } G \text{ is not a number}. \end{cases} \tag{8.2}$$

Exercise 8.4. Compute the left and right stops of each position marked with a box below. Rest assured that the game is in canonical form, so none of the interior nodes are numbers in disguise:

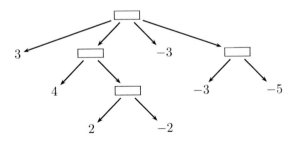

The main importance in knowing the adorned left and right stops of a game is that it allows us to determine precisely how G compares to any number x.

Theorem 8.5. *If G is a short game and x is a number, then*

- *if the value of the left stop of G is smaller than x, then $G < x$;*

- *if the value of the left stop of G is equal to x, then $G \leq x$ when $\mathbf{LS}(G) = x_R$ while G is confused with or greater than x when $\mathbf{LS}(G) = x_L$;*

- *if the value of the left stop of G is greater than x, then G is confused with or greater than x.*

Proof: Suppose that the value of the left stop of G is $y < x$ and consider the game $G - x$. Since the value of the right stop of G is at most the value of its left stop (as was shown in Theorem 5.34), Right has an option G^R with $\mathbf{LS}(G^R) \leq y$. By induction $G^R - x < 0$, so the option $G^R - x$ is a winning one for Right from $G - x$. Suppose that Left were to move first in $G - x$. A move in $-x$ is simply bad, so consider options of the form $G^L - x$. By the definition of the left stop, either $G^L - x = y - x < 0$ or the right stopping value of G^L is less than or equal to y. In that case, Right has an option to $G^{LR} - x$, where G^{LR} has left stopping value less than or equal to y, and so again by induction $G^{LR} - x < 0$. Thus, we see that Right has a winning strategy in $G - x$ moving either first or second, and hence $G - x < 0$.

Now suppose that $\mathbf{LS}(G) = x_R$ and consider again the game $G - x$. We must show that Right has a winning strategy moving second. Again, we need only consider options of the form $G^L - x$. If the right stopping value of G^L is smaller than x, then we continue as above. The only other possible value is x itself, and in this case it must also be true that $\mathbf{RS}(G^L) = x_R$ (since otherwise, $\mathbf{LS}(G) = x_L$, not x_R). So, Right can make a move to a position whose left stop is x_R and induction applies.

Now suppose that $\mathbf{LS}(G) \geq x_L$ (this includes the case where the left stopping value of G is greater than x). Then, moving first, Left can win in $G - x$ by choosing a left option G^L with $\mathbf{RS}(G^L) \geq x_L$ and applying the already proved results (with left and right interchanged). □

Combining this proposition with the dual one in which the roles of Left and Right are interchanged does indeed allow us to find how any short game G compares to any number. Specifically, associated to every game H there is a *confusion interval*, the set of numbers x with which H is confused. This confusion interval can be computed directly from the stops of the game as follows. The confusion interval of H is

$$\begin{array}{ll} [a,b] & \text{exactly if } \mathbf{LS}(H) = b_L \text{ and } \mathbf{RS}(H) = a_R; \\ (a,b] & \text{exactly if } \mathbf{LS}(H) = b_L \text{ and } \mathbf{RS}(H) = a_L; \\ [a,b) & \text{exactly if } \mathbf{LS}(H) = b_R \text{ and } \mathbf{RS}(H) = a_R; \\ (a,b) & \text{exactly if } \mathbf{LS}(H) = b_R \text{ and } \mathbf{RS}(H) = a_L. \end{array}$$

Notice that if G is a number plus an infinitesimal then strange intervals can occur. For example, the confusion interval of 0 is $(0,0)$ — no number is confused with 0. However, the confusion interval of $*$ is $[0,0]$ — 0 is the only number confused with $*$. For $\uparrow$, $\mathbf{LS}(\uparrow) = 0_R$ and $\mathbf{RS}(\uparrow) = 0_R$, and the confusion interval is $[0,0)$. Similarly, the confusion interval for $\downarrow$ is $(0,0]$. These are occasionally useful for induction purposes, but in practice only confusion intervals of hot games are of interest.

Let's see these principles in action and check them with an example. Consider the game $G = \{* \mid -1\}$ for which we know $\mathbf{LS}(G) = 0_R$ and $\mathbf{RS}(G) = -1_R$. Certainly, if $x > 0$, $G < x$ by the previous discussion, and if $x < -1$, $G > x$ by the dual form. For $-1 < x < 0$ Theorem 8.5 (and its dual) guarantees that G is both confused with or greater than x and confused with or less than x. So, for such x, G must be confused with x. What about $x = -1$? Again, the two conditions "confused with or greater than" and "confused with or less than" both apply, so G is confused with -1. For safety we can certainly check this directly. In $G + 1$ if Right moves first he moves to $-1 + 1 = 0$ and wins, while if Left moves first she moves to $* + 1$ and wins. Finally, how does G compare with 0, its left stop? This time we know that $G \leq 0$ and also that G is confused with or less than 0. So, the only possibility is that $G < 0$. To summarize, G has a confusion interval of $[-1, 0)$ and is greater than all numbers below its confusion interval and less than all numbers greater than its confusion interval.

8.2 Coping with Confusion

A game whose confusion interval has positive length is in a state of some excitement. Each player is eager to make the first move as this guarantees a better outcome for him or her than if his or her opponent moves first. The value of the game can be thought of as an indeterminate cloud, covering the confusion interval. In such a situation it is certainly natural to ask, "Is there a fair settlement about the value of the game?" If such a settlement — let us call it the *mean value* of the game — exists, then certainly it should lie in the confusion interval. In some cases it is easy to propose the value of such a settlement.

Consider first the simple switch $G = \{b \mid a\}$ (with $b > a$). If Left moves first the value is b, while if Right moves first the value is a. In fact, the sum

$$\{b \mid a\} + \{b \mid a\} = a + b.$$

So, playing G twice (in some sense) gives a value of $a + b$. In fact, $2k$ copies of G is exactly $k(a + b)$, while $2k + 1$ copies of G equal $k(a + b) + G$, whose confusion interval has width only $b - a$. Therefore, the mean value of G should be $\frac{(a+b)}{2}$.

Is it then the case that the mean value of a game should simply be the midpoint of its confusion interval? Unfortunately (?) not. Consider the game

$$H = \left\{ 19 \mid 1 \;\middle\|\; -1 \right\} = \{10 + \{9 \mid -9\} \mid -1\}.$$

Certainly $\mathbf{LS}(H) = 1_R$ and $\mathbf{RS}(H) = -1_R$, so the confusion interval is $[-1, 1)$. Is the mean value 0? No. Consider first $H + H$ and let us work out the stops of this game. Left's move is to

$$10 + \{9 \mid -9\} + H.$$

Right could now choose to move in H (to -1), and Left would collect 9 more stopping at 18. It would be better for Right to move in $\{9 \mid -9\}$,leaving $1 + H$ with Left to move and eventually a left stop of 2_R. What about the right stop? Right's first move is to $H - 1$, and from there things proceed automatically to 0 with Left to move, so the right stop is 0_R. Thus, the confusion interval of $H + H$ is $[0, 2)$. Continuing this analysis we see that whenever Left makes a move in a copy of H, Right's reply (to minimize the stopping value) will be in $\{9 \mid -9\}$. So basically, once Left gets a move in, she picks up 1 point in each remaining copy of H. If we play n copies of H, the right stopping value will be $n - 2$ and the left stopping value n. So, the mean value of H must be 1 — the very edge of the confusion interval.

Left's move in H (or sums of copies of H) to $10 + \{9 \mid -9\}$ is an embodiment of the concept of a *point in sente*. It is a move that creates a threat to gain a large amount (by following up with a move to 9). This threat dominates all other considerations and must be answered by Right. As a result, in the exchange of these two moves, Left gains a "point" while maintaining the initiative.

We are left with two problems. First of all, is there a well-defined notion of "mean value"? Secondly, how can we compute it in practice?

We can provide an affirmative answer to the first question now, while answering the second will occupy us for the rest of the chapter. Let $n \cdot G$ denote the sum of n copies of G (if n is a non-negative integer) or of $-n$ copies of $-G$ (if n is a negative integer).[1]

Theorem 8.6. (Mean Value) *For every short game G, there is a number $m(G)$ (the mean value of G) and a number t such that*

$$n \cdot m(G) - t \le n \cdot G \le n \cdot m(G) + t$$

for all integers n.

Proof: We begin with an observation about stopping values. Suppose that G and H are any two short games. Then,[2]

---

[1] The probabilist might compare this theorem with the notions of mean and standard deviation from probability. Whereas the standard deviation of a sum of n independent, identically distributed random variables grows by a factor of $\sqrt{n}$, the temperature (defined on page 189) of a sum of games remains bounded.

[2] If $\mathbf{LS}(G) = 2_L$ and $\mathbf{RS}(G) = 4_R$, then $\mathbf{LS}(G + H)$ could be 6_L or 6_R, so we are not adorning $\mathbf{LS}(G + H) = 6$.

$$\mathbf{RS}(G) + \mathbf{RS}(H) \leq \mathbf{RS}(G + H)$$
$$\leq \mathbf{RS}(G) + \mathbf{LS}(H)$$
$$\leq \mathbf{LS}(G + H)$$
$$\leq \mathbf{LS}(G) + \mathbf{LS}(H).$$

All of these inequalities follow from the first one after suitable changes of sign. And the first one is easy, since Left, playing second in $G + H$, can guarantee a stopping value greater than or equal to $\mathbf{RS}(G) + \mathbf{RS}(H)$ simply by answering Right's move in whichever game he happens to play.

Now consider $\mathbf{RS}(n \cdot G)$ (for simplicity assume that n is a positive integer). Certainly $\mathbf{RS}(n \cdot G) \leq \mathbf{LS}(n \cdot G)$. Of course, they are equal if $n \cdot G$ happens to be a number. Otherwise, $\mathbf{LS}(n \cdot G) = \mathbf{RS}((n-1) \cdot G + G^L)$ for some left option G^L of G. Since $(n-1) \cdot G + G^L = n \cdot G + (G^L - G)$, our observation above shows that

$$\mathbf{RS}((n-1) \cdot G + G^L) \leq \mathbf{RS}(n \cdot G) + \mathbf{LS}(G^L - G).$$

So, $\mathbf{LS}(n \cdot G)$ is sandwiched between $\mathbf{RS}(n \cdot G)$ and $\mathbf{RS}(n \cdot G) + \mathbf{LS}(G^L - G)$. In particular,

$$\mathbf{LS}(n \cdot G) - \mathbf{RS}(n \cdot G) \leq \mathbf{LS}(G^L - G).$$

That is, the width of the confusion interval of $n \cdot G$ is bounded. We also know that

$$\mathbf{LS}((k + m) \cdot G) \leq \mathbf{LS}(k \cdot G) + \mathbf{LS}(m \cdot G).$$

This means that the sequence $\mathbf{LS}(n \cdot G)$ is a *subadditive* sequence. It is well known that for any subadditive sequence a_n, the limit as n tends to infinity of $\frac{a_n}{n}$ exists and is either $-\infty$ or the lower bound of $\frac{a_n}{n}$. But dually in our case, the sequence $\mathbf{RS}(n \cdot G)$ is *superadditive* and $\mathbf{RS}(n \cdot G)/n$ has a limit that is either ∞ or the upper bound of this sequence. However, $\mathbf{RS}(n \cdot G) \leq \mathbf{LS}(n \cdot G)$ and their difference is bounded, so both the limits must actually exist and be a single common number, which we dub $m(G)$.

Now since the width of the confusion intervals for $n \cdot G$ are bounded, and $\mathbf{RS}(n \cdot G) \leq nm(G) \leq \mathbf{LS}(n \cdot G)$ for all n, there is a t such that for all n,

$$nm(G) - t < \mathbf{RS}(n \cdot G) \leq \mathbf{LS}(n \cdot G) < nm(G) + t,$$

and so in particular

$$n\,m(G) - t \leq n \cdot G \leq n\,m(G) + t,$$

as claimed. □

If we believe that the mean value $m(G)$ represents a fair value for G as a number, then we would hope that $m(G + H) = m(G) + m(H)$. Indeed this is the case.

Theorem 8.7. *For any two games G and H, $m(G + H) = m(G) + m(H)$.*

Proof: We know that

$$\mathbf{RS}(n \cdot G) + \mathbf{RS}(n \cdot H) \leq \mathbf{RS}(n \cdot (G + H)) \leq \mathbf{LS}(n \cdot (G + H))$$
$$\leq \mathbf{LS}(n \cdot G) + \mathbf{LS}(n \cdot H).$$

If we divide by n and then take a limit as n tends to infinity, we get

$$m(G) + m(H) \leq m(G + H) \leq m(G) + m(H),$$

exactly as we wanted.   □

Exercise 8.8. In SUBTRACTION$(1, 2 \mid 2, 3)$, what is the approximate value of the disjunctive sum of 6 heaps of size 10, 20 of size 11, and 132 of size 31?

8.3   Cooling Things Down

The next problem is how to compute the value $m(G)$, as well as possibly obtain extra information about G, which can be useful in play. The proofs in the preceding section give us a method in principle for computing $m(G)$, but in practice they are not necessarily so useful. Unless we are in luck and can see some obvious pattern in the canonical forms of the games $n \cdot G$ (for example, if $4 \cdot G$ turns out to be a number as was the case in one of our preceding examples), then computing the limits required to evaluate $m(G)$ will be impracticable. The technique we require, which allows an efficient computation of $m(G)$, is called *cooling,* and to introduce it we can do no better than quote verbatim from *ONAG* [Con01, p. 102]:

> We can regard the game G as vibrating between its Right and Left values in such a way that on average its center of mass is at $m(G)$. So in order to compute $m(G)$ we must find some way of cooling it down so as to quench these vibrations, and perhaps if we cool it sufficiently far, it will cease to vibrate at all, and *freeze* at $m(G)$.

Consider the simplest sort of hot game, a switch such as $\pm 2 = \{2 \mid -2\}$. As both players are keen to get the opportunity to move first, they will (presumably) not object if we charge them some fraction of a move for the privilege. Suppose, for example, that we asked Left for a donation of one move to Right in exchange for the privilege of moving first. She would certainly agree to accept the offer since her move would still be to $2 - 1 = 1$, a position favorable to her. If we made the corresponding offer to Right, he too would be happy to accept. However, suppose that we asked for a donation of three moves to Right in exchange for the first move. Now Left would scornfully decline, since we are giving her the opportunity to move to $2 - 3 = -1$.

The matched pair of offers "you can move first in G if you donate t moves to your opponent" represents an attempt to reduce the heat of battle. For the moment we will say that this represents cooling G by t. (We will see shortly that to define this operation precisely we need to be a little more careful.) For example, if we play the sum of ± 2 and the same game cooled by $\frac{1}{2}$, then because of the reduced incentive to play in the latter component, the first move will occur in ± 2.

Now instead of ± 2 let's look at the game $\{4 \mid 2\}$. If we adopt the naive definition of cooling suggested above, then what happens when we cool the game by, say, 10? We would get $\{-6 \mid 12\} = 0$. However, if we cool by exactly 1 we would get $3*$, and if we cool by anything just slightly larger than 1 we would get 3 (and two players who were unhappy about accepting the offer). Once a game has been cooled enough so that it is infinitesimally close to a number, neither player will be willing to play if we continue to cool it further — we recognize this by saying that the game has been *frozen* and agree to declare at this point that cooling it further has no effect. We also define the *temperature* of G to be the smallest amount of cooling required to freeze it (in other words, we think of all games having the same "freezing point"). We denote the result of cooling a game G by t as G_t. So,

$$(\{4 \mid 2\})_t = \begin{cases} \{4 - t \mid 2 + t\} & \text{if } t < 1, \\ \{3 \mid 3\} = 3* & \text{if } t = 1, \\ 3 & \text{if } t > 1. \end{cases}$$

Consider now the game $g = \{\{4 \mid 2\} \mid -2\}$. In isolation, g does not look very different from ± 2. After all, if Left moves first then after Right's reply she'll be two moves ahead, while if Right moves first he will wind up two moves ahead. However, the fact that Right needs to make an extra move in order to restrict Left's gains (in other words, the fact that Left's option was to a hot game rather than a number) has a significant effect on matters when we play sums of g with other games. For example, while $\pm 2 + \pm 2 = 0$, $g + g = \{2 \mid 0\}$ and $g + g + g + g = 2$. This last equation in particular implies that the mean value of g is $\frac{1}{2}$. Can we capture the mean value by looking at the results of cooling g by various amounts? Since cooling is a universal process, when we cool g we must also cool any of its unfrozen options. So, for small values of t (we will see how small in a moment),

$$g_t = \{-t + \{4 - t \mid 2 + t\} \mid -2 + t\} = \{\{4 - 2t \mid 2\} \mid -2 + t\}.$$

For $t < 1$ the left option is still hot and the right option is certainly smaller, so this equation holds for such t. For $t = 1$ the left option becomes $2*$ and

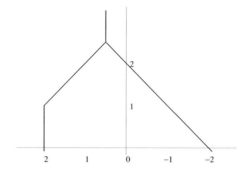

Figure 8.1. Thermograph of $g = \{4 \mid 2 \parallel -2\}$.

thereafter freezes. So, for $t = 1 + s$ (and s reasonably small),

$$g_t = \{-s + 2 \mid -1 + s\}.$$

Now things finally freeze at $s = \frac{3}{2}$, i.e., $t = \frac{5}{2}$, with

$$g_{\frac{5}{2}} = \left(\frac{1}{2}\right)*.$$

Therefore, the frozen value of g is $\frac{1}{2}$ exactly matching its mean value. Further-more, the temperature of g is $\frac{5}{2}$.

The results of cooling g by various amounts are conveniently plotted on its *thermograph*. In the thermograph, at height t we plot both $\mathbf{LS}(g_t)$ and $\mathbf{RS}(g_t)$. For historical reasons, the horizontal axis of a thermograph is reversed from the standard orientation. That is, values more favorable to Left (i.e., larger values) are placed at the left-hand side, rather than the right. The thermograph of g is shown in Figure 8.1.

In order to prove properties about cooling, it turns out to be simplest (arguably!) to first define the thermograph of a game inductively, and then to derive properties about cooling from that. The thermograph of any game will be a set of points in the plane. For each value of $t \geq 0$, the set will contain two points, which may coincide: $(\mathbf{LS}(G_t), t)$ and $(\mathbf{RS}(G_t), t)$. The notation that we use here for these points will be justified post facto. The set of points $(\mathbf{LS}(G_t), t)$ for $t \geq 0$ will form a path whose segments are either vertical or slanting on a 45 degree angle upward and to the right, finishing in a vertical mast. Likewise, the set of points $(\mathbf{RS}(G_t), t)$ for $t \geq 0$ will form a path whose segments are either vertical or slanting on a 45 degree angle upward and to the left, finishing in the same vertical mast.

So, let G be a game, and suppose that we have access to the thermographs of all the options of G, and that these have the characteristics we have mentioned. We define the thermograph of G as follows:

- If G is a number x, then $(\mathbf{LS}(G_t), t) = (\mathbf{RS}(G_t), t) = (x, t)$ for all t.

- Otherwise, take $\mathbf{LS}(G_t)$ to be the maximum of $\mathbf{RS}(G_t^L) - t$ and $\mathbf{RS}(G_t)$ to be the minimum of $\mathbf{LS}(G_t^R) + t$ *unless* this would require $\mathbf{LS}(G_t) < \mathbf{RS}(G_t)$. In that case, $\mathbf{LS}(G_t) = \mathbf{RS}(G_t) = \mathbf{LS}(G_u)$ where u is the smallest value for which $\mathbf{LS}(G_u) = \mathbf{RS}(G_u)$.

Such an involved definition needs to be checked out in order to ensure that it works properly. The first part is simple enough (and clearly satisfies the characteristics we have specified). So, suppose that the second part of the definition is in force. Under the assumptions made, all the segments $\mathbf{RS}(G_t^L)$ are vertical or upward left. When we subtract t from them, we get segments that are upward right or vertical (remember that the axis is reversed from its conventional orientation). This property is preserved by taking the maximum. The same remark applies to the $\mathbf{LS}(G_t^R) + t$. So that gives us two segments, each finishing in a diagonal. These segments intersect for the first time at some height, and the remainder of the definition says that we form the thermograph of G at and above that height, by finishing with a vertical mast. The horizontal coordinate defined by this mast is called the *mast value* of G.

Now let us try and define how to cool a game G by t in order to obtain G_t. We set inductively

$$G_t = \left\{ G_t^L - t \mid G_t^R + t \right\}$$

unless there is some $u < t$ where G_u has the property that

$$\mathbf{LS}(G_u) = \mathbf{RS}(G_u) = x,$$

in which case we define $G_t = x$.

This definition works precisely because our consideration of the thermograph (which we can see plots the left and right stops of G_t) guarantees that there will be a smallest such u.

We will try to consider these two definitions in parallel with respect to the following example:

$$G = \{2, \{4 \mid 1\} \mid \{-1 \mid -2\}\}.$$

First, we superimpose the thermographs of 2 and $\{4 \mid 1\}$ and highlight the *leftmost right boundary* (see Figure 8.2).

That boundary, shifted rightward by t units at height t, will form the left boundary of the final thermograph, up to the point where it intersects with the right boundary. The right boundary is simply obtained from the left boundary of the thermograph of $\{-1 \mid -2\}$ shifted leftward by t units at height t. The finished thermograph is shown as the highlighted central part in Figure 8.3.

Now let us consider the inductive definition of cooling G by t. For sufficiently small t,

$$G_t = \{2 - t, \{4 - t \mid 1 + t\} - t \mid \{-1 - t \mid -2 + t\} + t\}.$$

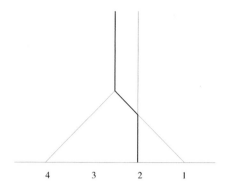

Figure 8.2. Thermographs of 2 and $\{4 \mid 1\}$ superimposed.

How small is "sufficiently small"? Exactly until some pair of left and right stops of G_t, or one of its subgames, coincides. By inspection, this occurs when $t = \frac{1}{2}$. For $t \leq \frac{1}{2}$, $\mathbf{LS}(G_t) = 2 - t$ and $\mathbf{RS}(G_t) = -1$. When we cool by more than $\frac{1}{2}$, the right-hand option of G freezes and the right stopping value becomes $-\frac{3}{2} + t$. The next important temperature is $t = 1$. At this point, the two left options of G_t are 1 and $\{3 \mid 1\}$. Up to this point, in computing $\mathbf{LS}(G_t)$, Left has preferred the move in the first option. However, for $t > 1$, Left will prefer to move in the second option as its right stop remains at 1, whereas the right stop of $2 - t$ becomes smaller than 1. At $t = \frac{3}{2}$ the hot left option freezes. Thereafter, the left stopping value becomes $\frac{5}{2} - t$. Finally, the left and right boundaries collide when $\frac{5}{2} - t = -\frac{3}{2} + t$ (that is, at $t = 2$), and the game freezes entirely at this point. Collecting this information, we obtain

$$
G_t = \begin{cases}
\{2 - t, \{4 - 2t \mid 1\} \mid \{-1 \mid -2 + 2t\}\} & \text{for } 0 \leq t \leq \frac{1}{2}, \\
\{2 - t, \{4 - 2t \mid 1\} \mid -\frac{3}{2} + t\} & \text{for } \frac{1}{2} < t \leq 1, \\
\{\{4 - 2t \mid 1\} \mid -\frac{3}{2} + t\} & \text{for } 1 < t \leq \frac{3}{2}, \\
\{\frac{5}{2} - t \mid -\frac{3}{2} + t\} & \text{for } \frac{3}{2} < t \leq 2, \\
\frac{1}{2} & \text{for } t > 2.
\end{cases}
$$

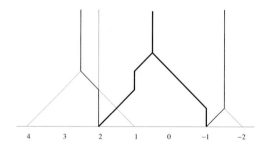

Figure 8.3. Constructing the thermograph of $\{2, \{4 \mid 1\} \mid \{-1 \mid -2\}\}$ from those of its options.

We see that the thermograph reflects exactly the left and right stopping values of G_t (as indeed it must!).

To continue with the development of the theory of thermography, we introduce one more notion. Any game G is really just a tree, whose root is the game itself and which has leftward pointing branches to the left options of G and rightward pointing branches to the right options. The leaves of such a tree are the terminal positions in which neither player can move. Ordinary induction on games can be viewed as an induction on these trees — based on an inductive hypothesis that the result holds for all subtrees of a given tree. Another tree that we can associate with a game is the *stopped tree*. In the stopped tree we recognize that when play has proceeded to a point where the value is a number, neither player will be terribly interested in continuing. So, instead of completing the tree, we simply label such a vertex with the number that is its value. Now the leaves of the tree simply carry numbers.

Given a stopped tree, it is trivial to work out the left and right stops of any of its internal vertices as the left and right stops of a number are the number itself, and otherwise we can use the usual inductive rule. Likewise, the stopped tree of G provides a simple means of computing (the stopped tree of) G_t. At each leaf v, count the excess of left branchings over right branchings when we follow the path from the root to that leaf. Call this number e_v. Now replace the label n_v on that leaf by $n_v - e_v x$. We will think of x as a variable temperature, which we will allow to run from 0 through t. In the simplest case, if we simply replaced x by t, this new tree would represent the value obtained when G_t is played out to this leaf. However, we know that cooling is not quite so simple and that for actual concrete values of t we must not allow overcooling.

So, we think of the cooling process as taking place in a series of phases. One type of phase will be called *congealing*. In a congealing phase, any internal vertex whose left and right stops are now equal is replaced by the number equal to those stops (note that this number might be an expression like $2 - x$). The other type of phase will simply be allowing the value of x to increase by some amount (actual cooling). A *critical temperature* is a value of x at which congealing can take place.

It is easy to check, though there is a *little* thought required, that any critical temperature is a dyadic rational.

Now, the point is that G_t is just the game that we get by alternately congealing and cooling, pausing for a congealing step each time we reach (and intend to exceed) a critical temperature. This is essentially exactly the description of the cooling process that we followed in the example where

$$G = \{2, \{4 \mid 1\} \mid \{-1 \mid -2\}\}.$$

From the description of the congealing process, it is clear that the stopped tree of G_t is always a subtree of the stopped tree of G for any temperature

$t \geq 0$. This gives us an inductive tool to prove the following critical results about cooling.

Lemma 8.9. *Let X and Y be games, and suppose that t is a sufficiently small real number such that all of X_t, Y_t, and $(X + Y)_t$ can be computed by the inductive definition. Then, $X_t + Y_t = (X + Y)_t$. If any of X, Y, or $X + Y$ is a number, then this result is also true.*

Exercise 8.10. Prove the lemma in the case that none of X, Y, and $X + Y$ is a number. (You are asked to prove the other cases in Problem 10.)

Theorem 8.11. *Cooling is an additive function on games. That is, for any games G and H, and any $t \geq 0$, $G_t + H_t = (G + H)_t$.*

Proof: Contrary to our usual practice, we will give this inductive proof in terms of a minimal counterexample argument. That is, suppose that the result were false. Then, there would be a counterexample $(G, H, G + H)$ with the sum of the sizes of the stopped trees of G, H, and $G + H$ minimized.

In this counterexample, none of the three parts can be a number by Lemma 8.9.

Consider starting the cooling process. If a congealing phase is needed, to begin with we obtain G', H', and $(G + H)'$. Do we know that $G' + H' = (G + H)'$? Yes we do, because congealing can be thought of as "cooling by an infinitesimal." For such cooling the inductive definition of the cooling operation is always applied and Lemma 8.9 shows that this is additive. However, we now know that $(G', H', G' + H')$ is another counterexample to the theorem, and the total size of the stopped trees has decreased.

Thus, we can cool at least until the smallest critical temperature of G, H, and $G + H$ without violating the theorem. However, this leaves us in exactly the same situation as above. That is, after cooling by this amount, we will congeal. At least one of the trees will become smaller, and we will still have a counterexample.   □

Finally, we are in a position to illustrate the connection between cooling and the mean value of a game.

Theorem 8.12. *The mast value and the mean value of a game G are the same.*

Proof: Let $M(G)$ denote the mast value of G. Then, for all sufficiently large t, $M(G) = G_t$. By the additivity of cooling, for all positive integers n and all sufficiently large t, $nM(G) = (n \cdot G)_t = M(n \cdot G)$. However, the mast value of any game lies between its left and right stops, so we have

$$\mathbf{RS}(n \cdot G) \leq nM(G) \leq \mathbf{LS}(n \cdot G).$$

Dividing this inequality by n and taking the limit as n tends to infinity gives

$$m(G) \leq M(G) \leq m(G);$$

that is, $m(G) = M(G)$ as claimed. □

The significance of this theorem is that the mast value of G can be easily computed simply by computing the thermograph of G, and the description above provides a simple and efficient algorithm for performing that computation.

Proposition 8.13. *Let G and H be games. The temperature of $G + H$ is less than or equal to the maximum of the temperatures of G and H.*

Proof: The temperature $t(X)$ of a game X is the least real number that has the property that, for all $t > t(X)$, X_t is a number. Suppose that $t > \max(t(G), t(H))$. Then, $(G + H)_t = G_t + H_t$, which is the sum of two numbers and so is a number. Thus, $t(G + H) \leq \max(t(G), t(H))$ as claimed. □

Note that it is certainly possible that $t(G + H) < \max(t(G), t(H))$ in the situation just described. The most obvious case is when $G = -H$. Problem 9 asks you to show that if $t(G) \neq t(H)$ then $t(G + H) = \max(t(G), t(H))$.

8.4   Strategies for Playing Hot Games

Thermographs are quite helpful for playing sums of hot games. Be warned, however, that different games can have the same thermograph. For example, $X = \pm 1$, $Y = \{1, 1 \pm 1 \mid -1, -1 \pm 1\}$, and even $X + Y$ all have the thermograph given by

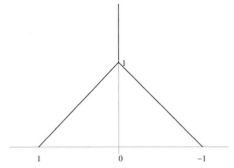

And yet, on $-1 + X + Y$, the only winning move by Left is on Y. Therefore, the thermograph alone cannot dictate optimal play. Phrased differently, canonical form is *the* information required to know how to play a game in any context; thermographs lose some of this information.

We already know from Chapter 5 that it is NP-hard to decide the value of the sum of games of the form $\{a \mid \{b \mid c\}\}$. Yet, if we know the thermographs of all components, they can be used to guide play. In other hot games (such as AMAZONS and GO), it is quite difficult, indeed, to compute the full thermograph of each component.[3] However, even in those cases, we can often estimate the thermograph or calculate the temperature (but not the thermograph). The central question is how should we now interpret the information given by temperature and thermography?

Here, we only give a brief description, a manual of "how-to" with appropriate warnings. For more on the real "how?" and "why?" the reader should consult [BCG01, ML06, Ber96].

Three main strategies that use temperature have been identified.[4] One of these strategies, Thermostrat, requires a little additional machinery:

Definition 8.14. Given the sum $A + B + C + \ldots$, let

$$W_t = \max\{W_t(A), W_t(B), W_t(C), \ldots\},$$

where $W_t(X) = \mathbf{LS}(X_t) - \mathbf{RS}(X_t)$ is the width of the thermograph of X at temperature t. The *compound thermograph* of the sum has right boundary $\mathbf{RS}(A_t) + \mathbf{RS}(B_t) + \mathbf{RS}(C_t) + \cdots$ and left boundary $\mathbf{RS}(A_t) + \mathbf{RS}(B_t) + \mathbf{RS}(C_t) + \cdots + W_t$. The *ambient temperature* of $A + B + C + \ldots$ is the least T for which $\mathbf{RS}(A_T) + \mathbf{RS}(B_T) + \mathbf{RS}(C_T) + \ldots + W_T$ is maximal. That is, the height of the lowest horizontal line whose leftmost intersection with the compound thermograph is farthest left.

While the underlying thermographs have slopes of 1, 0, and -1, compound thermographs can have integral slopes. (Slopes are viewed with the thermograph turned 90 degrees, for temperature is the independent variable.)

There are three standard strategies based on temperature:

Hotstrat: Move in a component whose temperature is maximal. While the strategy is simple and intuitive, there are no theorems guaranteeing good performance.

Thermostrat: Find the ambient temperature T and move in the component whose thermograph is widest at T. This strategy is guaranteed to achieve an outcome whose value differs from the mean by at most T; in fact, *WW* [BCG01] uses Thermostrat to give a constructive proof of Theorem 8.12. This strategy is used when the thermographs of each component can be easily calculated.

---

[3]In computer analysis of games, a more urgent question, one we do not even mention, is to first identify the components of the sum, and this is not necessarily easy.

[4]Computer scientists have many other strategies at their disposal for identifying reasonable moves. Here, we are concerned with provably good strategies for playing in disjoint sums of games.

Sentestrat: This time, define the *ambient temperature* as the lowest temperature of any of your previous moves; it starts out as infinite. If your opponent has just moved in a component leaving it at a temperature above the ambient one, respond in that component; otherwise play Hotstrat. Sentestrat is guaranteed to do well in the context of enough switch games $\pm x$ for numbers x. Since Sentestrat requires only the temperature of each component (and not its full thermograph), it is widely applicable and should most often be the preferred strategy.

Exercise 8.15. Consider $P + Q$, where $P = \{\{20 \mid 0\} \mid -1\}$ and $Q = \{2 \mid -2\}$.

(a) Does Hotstrat advise optimal play when Left moves first? When Right moves first?

(b) Answer the same questions for Thermostrat.

Exercise 8.16. There is asymmetry in the description of Sentestrat, for it defines the ambient temperature to be the lowest temperature of any of *your* previous moves (and not your opponent's). Why does this make sense?

8.5   Norton Products

We understand numbers and many infinitesimals, but we do not understand many hot games. Hence, we often attempt to associate the hot positions from a particular game with numbers or tepid games in a one-to-one fashion, in the hopes of gaining insight into the hot game. (Recall from Definition 6.17 on page 147 that a tepid game is the sum of a number and a non-zero infinitesimal.)

But cooling by t is a many-to-one function. For instance, all infinitesimals cool to 0 when cooled by any $x > 0$. So, there are many ways to add heat to a game in a way that inverts cooling. We will focus on one such way, the *Norton product,* since it is guaranteed to be linear and order-preserving. The serious practitioner will also want to study more general ways to invert cooling such as the *heating* and *overheating* operators defined in *WW* [BCG01, pp. 167–178]. Like Norton products, these can also provide a terse representation of an otherwise complex game. Although well behaved in practice, heating and overheating are not a priori guaranteed to be linear.

Definition 8.17. Let G be any game, let *unit* U be any positive game in canonical form, and let
$$\tau = U + \mathcal{I},$$

where $\mathcal{I}$ are the incentives from U. Then, the *Norton product* of G by U is given by

$$G \cdot U = \begin{cases} \overbrace{U + U + U + \cdots + U}^{n} & \text{if } G = n \text{ is a positive integer,} \\ 0 & \text{if } G = 0, \\ \overbrace{-U - U - U - \cdots - U}^{n} & \text{if } G = -n \text{ is a negative integer,} \\ \left\{ \mathcal{G}^L \cdot U + \tau \mid \mathcal{G}^R \cdot U - \tau \right\} & \text{otherwise.} \end{cases}$$

Note that while $\mathcal{I}$ and therefore τ are both sets of games, in most applications τ is a singleton. We will refer to a typical element of τ by t, and $I_t \in \mathcal{I}$ is the incentive corresponding to t in τ.

Example 8.18. Define the unit $U = x*$, where x is a number. In this case, the left and right incentives are both $*$, and there is only one game $t \in \tau$, that being $t = x* + * = x$. So, we get

$$n \cdot x* = \begin{cases} nx & \text{if } n \text{ is even,} \\ nx* & \text{if } n \text{ is odd;} \end{cases}$$

$$* \cdot x* = \{0 + t \mid 0 - t\} = \{x \mid -x\} ;$$

$$\uparrow \cdot x* = \{0 + t \mid \{t \mid -t\} - t\} = \left\{ x \,\middle\|\, 0 \mid -2x \right\};$$

$$\frac{1}{2} \cdot x* = \{0 + t \mid t* - t\} = \{x \mid *\} .$$

Norton product by $1*$ appears prominently in one-point GO endgames [BW94]. Other cases of Norton products by $U = x*$ also appear. Norton product by $U = \uparrow$ is of special importance, and we will explore its use in Chapter 9.

Exercise 8.19. List $\frac{1}{2^n} \cdot 1*$ for integers $n \geq 0$ until the pattern is clear.

Ignoring base cases, when cooling by t each player is taxed t for moving. In a Norton product, each player benefits by t for each move. To show that Norton products are well behaved, we will echo the proof technique that we last saw in Lemma 5.16 on page 109.

Lemma 8.20. *For unit $U > 0$,*

$$A + B + C \geq 0 \iff A \cdot U + B \cdot U + C \cdot U \geq 0 \text{ and}$$
$$A + B + C \leq 0 \iff A \cdot U + B \cdot U + C \cdot U \leq 0.$$

Proof: In essence, play on $A + B + C$ mirrors play on $A \cdot U + B \cdot U + C \cdot U$ except that, on each play, a quantity $t \geq 0$ changes sides (or, more exactly, quantities τ change sides). Problem 12 asks you to prove that $t \geq 0$.

It suffices to prove the following stronger pair of assertions:

$$\text{If } A + B + C \rhd 0 \text{ then } A \cdot U + B \cdot U + C \cdot U - t \ \rhd \ 0. \qquad (8.3)$$
$$\text{If } A + B + C \geq 0 \text{ then } A \cdot U + B \cdot U + C \cdot U \geq 0. \qquad (8.4)$$

We could equally well have also proved the symmetric pair of assertions with $\rhd$ replaced by $\lhd$ and $\geq$ replaced by $\leq$, and together the four resulting "If ... then ..." statements imply the lemma.

We mean for (8.3) to hold for every $t \in \tau$. Now (8.3) and (8.4) do yield the first $\Longleftrightarrow$, since for every $t \geq 0$ (8.3) implies, "if $A + B + C \ \rhd \ 0$ then $A \cdot U + B \cdot U + C \cdot U \ \rhd \ 0$." By symmetry, we can change $\rhd$ to $\lhd$ in both places; then, taking the contrapositive yields, "if $A \cdot U + B \cdot U + C \cdot U \geq 0$ then $A + B + C \geq 0$."

We will now proceed to prove both (8.3) and (8.4) in tandem by induction.

To prove the assertion given in (8.3), assume that $A + B + C \ \rhd \ 0$ and so some $A^L + B + C \geq 0$. By the Number-Avoidance Theorem, we may assume that players move on non-integers when available. So, without loss of generality, either A is a non-integer or A, B, and C are all integers. If A is a non-integer, then Left has the same move available from $A \cdot U + B \cdot U + C \cdot U - t$ to $A^L \cdot U + B \cdot U + C \cdot U$, which she wins moving second by induction using (8.4). If, however, all three are integers, then $A + B + C$ is some integer $n \geq 1$, and Left wins moving first on n copies of the positive game U. Note that $n \cdot U - t = (n - 1) \cdot U - I_t$, which Left wins going first since $n - 1 \geq 0$, $U > 0$, and $I_t \ \rhd \ 0$.

For (8.4), if A is a non-integer then Right's move from $A \cdot U + B \cdot U + C \cdot U$ to $A^R \cdot U + B \cdot U + C \cdot U - t$ loses by induction using (8.3) since $A^R + B + C \ \rhd \ 0$. If, on the other hand, A is an integer, then if $A > 0$ Right's move is to some

$$(A - 1) \cdot U + U^R + B \cdot U + C \cdot U = (U^R - U) + A \cdot U + B \cdot U + C \cdot U.$$

If $A < 0$, Right's move is to some

$$(A + 1) \cdot U - U^L + B \cdot U + C \cdot U = (U - U^L) + A \cdot U + B \cdot U + C \cdot U.$$

Since $U^R - U$ and $U - U^L$ are negative incentives, these each equal some

$$(A + 1) \cdot U + B \cdot U + C \cdot U - t$$

for some $t \in \tau$. Now, $(A + 1) + B + C > A + B + C \geq 0$, and so by induction $(A + 1) \cdot U + B \cdot U + C \cdot U - t \ \rhd \ 0$. $\qquad \square$

Exercise 8.21. The last inductive step appears backward since it uses a statement about $A + 1$ to prove one about A. Explain why the proof is just fine.

Theorem 8.22. *The Norton product satisfies the following:*

Independence of form: If $A = B$ then $A \cdot U = B \cdot U$.

Monotonicity: $A \geq B$ if and only if $A \cdot U \geq B \cdot U$.

Distributivity: $(A + B) \cdot U = A \cdot U + B \cdot U$.

Proof: All three facts follow directly from Lemma 8.20. □

8.6 Domineering Snakes

Some snaky DOMINEERING repeating patterns are easily described by Norton product by $\frac{1}{2}*$ [Wol93]. First, we introduce some handy pictorial notation:

The little notches show how pieces fit together, and the repeated piece is shown with an integer indicating how many copies of the piece are required to form the snake.

The following table summarizes the values that appear when playing on these two snakes:

k	$\int g \stackrel{\text{def}}{=} g \cdot \frac{1}{2}*$					
0	$\int -2$	$*$	$1\|0$	$\int -2$	$*$	$1\|0$
1	0	$\int \frac{1}{2}$	$\int 2$	0	$\int \frac{3}{2}$	$\int 2$
2	$*$	$*\int \frac{5}{4}$	$*\int 2$	$*\int 2$	$*\int \frac{11}{4}$	$*\int 4$
3	$\int 1$	$\int 2$	$\int \frac{5}{2}$	$\int 3$	$\int 4$	$\int \frac{11}{2}$
4	$*\int \frac{3}{2}$	$*\int \frac{5}{2}$	$*\int \frac{13}{4}$	$*\int \frac{9}{2}$	$*\int \frac{11}{2}$	$*\int \frac{27}{4}$
5	$\int 2$	$\int 3$	$\int 4$	$\int 6$	$\int 7$	$\int 8$
6	$*\int 3$	$*\int \frac{7}{2}$	$*\int \frac{9}{2}$	$*\int 7$	$*\int \frac{17}{2}$	$*\int \frac{19}{2}$
	period 5, saltus $*\int 3 = \frac{3}{2}$			period 5, saltus $*\int 7 = \frac{7}{2}$		

When reading the table, exceptions to the period information are shown above a horizontal line. A few entries are omitted, since appropriately tacking on

tends to add 1 or $\int 2 = 2 \cdot \frac{1}{2}*$:

$$\lefthalf = \square + \square \quad \textit{except} \quad \lefthalf = \frac{1}{2}.$$

As an example of how to use the table, consider the position we saw at the start of the section:

This position appears in the 2nd column with $k = 3$ and so has value $2 \cdot \frac{1}{2}* = 1$. Were it the case that $k = 8$, the value would be

$$\int 2 + * \int 3 = * \int 5 = *+5 \cdot \frac{1}{2}* = \frac{5}{2}.$$

Proving that the table is correct is mechanical but quite tedious. For example, suppose that we wish to confirm the entry

for $k \equiv 3 \pmod 5$. Thus, suppose that $k = 5(j+1) + 3$ for some integer $j \geq 0$. We need to describe all the possible left and right options from this position, which are typically sums of pairs of entries from the table plus a multiple of the saltus. Assuming that the values provided for the options are correct, by induction we can confirm that the value provided for the original position is correct. Perhaps the proof is best left to the computer.

Exercise 8.23. Evaluate the following DOMINEERING position:

(Problem 15 asks you to find all winning moves.)

Problems

1. In TOPPLING DOMINOES find the canonical forms, means, and temperatures for

$$A = \text{▐▐▐▐▐▐} \, , \; B = \text{▐▐▐▐▐▐▐} \, , \; \text{and} \; C = \text{▐▐▐▐▐▐▐▐▐}$$

What are the left and right stops for $A + B + C$?

2. The canonical form of $A + B + C$ from Problem 1 is given by

$$\left\{ \begin{array}{c} \{\frac{7}{2} \mid 2 \parallel 1 \mid -\frac{1}{2}\}, \\ \{\frac{7}{2} \mid 2 \parallel\parallel 1 \mid \frac{1}{2} \parallel -\frac{1}{2} \mid -1\} \end{array} \middle| \begin{array}{c} \{1 \mid \frac{-1}{2} \parallel\parallel -\frac{3}{2} \mid -2 \parallel -3 \mid -\frac{7}{2}\}, \\ \{1 \mid \frac{1}{2} \parallel -\frac{1}{2} \mid -1 \parallel\parallel -\frac{3}{2} \mid -2 \parallel -3 \mid -\frac{7}{2}\} \end{array} \right\}.$$

Find the thermograph of $A + B + C$. (Do not be put off by the size of the canonical form; some positions appear more than once.) While you should show your work, feel free to check your answer in CGSuite using, for instance, `Plot(Thermograph({1|-1/2}))`.

3. In TOPPLING DOMINOES, with

$$G = \text{▐▐▐▐▐▐▐▐} \quad \text{and} \quad H = \text{▐▐▐▐▐▐}$$

find the canonical forms, stops, and thermographs for G, H, and $G + H$.

4. Find the stops, confusion interval, mean value, and temperature for the AMAZONS position

5. (Open ended) Practice calculating means and temperatures of hot games until you become proficient. Begin with games of the form $\{a \parallel b \mid c\}$. See what happens when you add an option. Use CGSuite to check your work. Try to find games of temperature $\frac{1}{2}$, $\frac{1}{4}$, $\frac{3}{8}$, etc.

Select two or three computations and write them up carefully so it is clear that you understand how to properly construct thermographs.

6. (a) Find a game with the following thermograph:

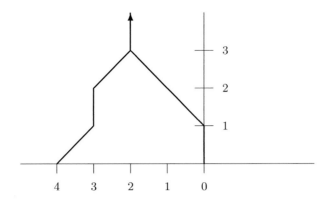

(b) How can *any* such game G, with the above thermograph, relate to zero? Be specific and explain your answer (i.e., can a game G with the above thermograph be greater than zero? Can another be less than zero? Equal? Incomparable?).

7. Find the stops and the confusion interval for AMAZONS positions of the form 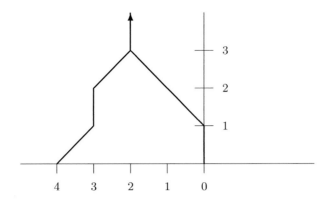 and show that the mean value is always $-\frac{1}{4}$.

8. The subtraction game SUBTRACTION$(1, 4, 10 \mid 4, 10, 13)$ is purely periodic with period 14. The first period is 0, 1, 2, 3, $\{3 \mid 0\}$, 1∗, 2∗, 3∗, 3, $\{3 \mid 1∗\}$, $\{2∗, \{3 \mid 1∗\} \mid 0\}$, $\{3∗ \mid 1\}$, $\{3 \mid 2\}$, $\{3, \{3 \mid 2\} \mid 0\}$. Suppose that you play a game G consisting of 50 heaps, 10 each of of sizes 10, 20, 30, 40, and 50. Determine numbers x and ϵ such that $x - \epsilon \leq G \leq x + \epsilon$ and prove your answer. How small can you make ϵ?

9. Show that if $t(G) \neq t(H)$ then $t(G + H) = \max(t(G), t(H))$.

10. Prove Lemma 8.9 in the case that at least one of X, Y, and $X + Y$ is a number.

11. Let U be a Norton unit, i.e., a positive game in canonical form.

   (a) For all games G, prove that if U is all-small, so is $G \cdot U$.

   (b) Prove that if U is infinitesimal, so is $G \cdot U$.

12. Let U be a Norton unit, i.e., a positive game in canonical form. Let $\tau = U + \mathcal{I}$ as in Definition 8.17 on page 197. Prove that all the values of τ are at least 0. (*Hint:* A typical $t \in \tau$ is either $U + U^L - U$ or $U + U - U^R$. You can use the fact that U has no reversible nor dominated options to show that each $t \geq 0$.)

13. In Problem 12, if $t = 0$, find, with proof, all possible values for U.

14. In the definition of Norton product, confirm that $U > 0$ is a necessary condition for all three assertions in Theorem 8.22. In particular, pretend that $U = *$. For each of the three assertions of the theorem, find games A and B violating the assertion. For your monotonicity and distributivity counterexamples, A and B should be in canonical form.

15. Find several winning moves for Left from the position in Exercise 8.23 on page 201. (There are seven winning moves in total.)

Preparation for Chapter 9

Prep Problem 9.1. Review reversibility and rederive the canonical form of $\Uparrow = \uparrow + \uparrow$ and of $\uparrow* = \uparrow + *$ from scratch. If you find yourself having to refer to Section 5.4, you can also try $\Uparrow*$.

To the instructor: The game of HACKENBUSH is ideal for covering integer atomic weights. Start with flower gardens from *WW* [BCG01, Ch. 7]. An advanced group of students will be unsatisfied without seeing proofs of atomic weights from *WW* [BCG01, Ch. 8].

Chapter 9
All-Small Games

It has long been an axiom of mine that the
little things are infinitely the most important.

Sir Arthur Conan Doyle in *The Adventures
of Sherlock Holmes: A Case of Identity*

If a game and all of its positions have the property that either both players have
a move or neither one does, then the game is called *all-small* (see page 119).
Typical all-small games are those, such as ALL-SMALL CLEAR THE POND, where
players are attempting to race their pieces off the board and which end when
one player has successfully done so, or games such as CLOBBER and CUTTHROAT
STARS where the objective is to eliminate all of your opponent's resources and
each of your moves must contribute to that elimination.

All-small games form a subset of the infinitesimal games and arise frequently as parts of positions in larger games. In practice, all-small games
are often quite subtle and challenging in part because complex canonical forms
abound, and it is difficult for the amateur to understand who has the advantage
and by how much in each local position.

In such games, it is often the case that many of the options are reasonably
close in value. By carefully establishing what we mean by *close,* we can define
an equivalence relationequivalence relation on the class of all-small games and
assign to each equivalence class a descriptive value called its *atomic weight.* The
atomic weight may be used as a close approximation of the game's real value
and, in many instances, determines the outcome class of the game and provides
a good move for the winner. While the results in this chapter are motivated by
the analysis of all-small games, many also extend to other infinitesimals.

9.1   Cast of Characters

Let us repeat the definition of an all-small game:

Definition 9.1. A game G is *all-small* if and only if either $G = \{ \mid \}$ or both $\mathcal{G}^L$ and $\mathcal{G}^R$ are non-empty and all the elements of $\mathcal{G}^L$ and $\mathcal{G}^R$ are all-small. Equivalently, every position of G is either terminal or has both left and right options.

As stated, whether or not a game is all-small depends on its form and not just its value. For instance, while $\{ \mid \}$ whose value is 0 is all-small, the game $\{-1 \mid 1\}$, which has the same value, is not all-small. In practice, this distinction is spurious since we are interested in playing games well — and playing games well depends only on knowing their values, not their forms. So, we will not pay any attention to this distinction in the sequel. Further justification for this decision can be found in Problem 12.

Any all-small game G is infinitesimal since Left can win $G + x$ for any positive number x by playing blindly in G as long as she has a move there. At some point either Right will have lost or Left will have no further options in G. In that case, though, Right will have no further options in G either since G is all-small. The remaining game will be some number $x' \geq x > 0$ (where inequality might occur if Right has foolishly ignored the Number-Avoidance Theorem and played on x at some point in the play). So, for any positive real number x, $G + x > 0$, and dually $G - x < 0$. However, not all infinitesimals are all-small; for example, $\mathbf{+}_2 = \{0 \mid \{0 \mid -2\}\}$ is not all-small, since from -2 Right has a move but Left does not. Nor, according to our comment above, is $_2$ equal to any all-small game since its canonical form is not all-small. We have seen that $\uparrow$, $*$, and $\uparrow\!*$ are all-small and so are all impartial games.

The game $\uparrow$ and the nimbers play central roles in the theory surrounding all-small games, so before attacking that theory headlong, we will do some warming up by considering properties of various games defined from them.

The game $\uparrow\!*n$

Our first warm-up routine will be to compute the canonical forms of sums of $\uparrow$ and nimbers.

Lemma 9.2. *Let n be a non-negative integer. Then, $\uparrow\!*n > 0$ if and only if $n \neq 1$.*

Proof: We have long since dealt with the cases $n = 0$ and $n = 1$. Suppose that $n > 1$ and consider $G = \uparrow\!*n$. Moving first, Left can move to $\uparrow$ and win. Right's options in G are: $\uparrow\!*n'$ for some $n' < n$ which is in either $\mathcal{L}$ or $\mathcal{N}$, and $*(n \oplus 1)$ which is also in $\mathcal{N}$. So, Left wins moving either first or second, and thus $\uparrow\!*n > 0$. $\qquad\square$

Corollary 9.3. *Let n and m be non-negative integers. Then $\uparrow\!*n > *m$ if and only if $m \neq n \oplus 1$.*

Exercise 9.4. Prove the corollary.

We already know the canonical forms:

$$\uparrow = \{0 \mid *\};$$
$$\uparrow\!* = \{0, * \mid 0\}.$$

The following lemma completes the catalog of canonical forms for the games $\uparrow\!*n$.

Lemma 9.5. *The canonical form of $\uparrow\!*n$ for $n > 1$ is $\{0 \mid *(n \oplus 1)\}$.*

Proof: To find the canonical form, we begin by listing the options:

$$\uparrow\!*n = \{*n, \uparrow, \uparrow\!*, \uparrow\!*2, \ldots, \uparrow\!*(n-1) \mid * + *n, \uparrow, \uparrow\!*, \uparrow\!*2, \ldots, \uparrow\!*(n-1)\}.$$

That is, Left's options are $*n$ and $\uparrow\!*m$ for $0 \leq m < n$; Right's options are $* + *n$ and $\uparrow\!*m$ for $0 \leq m < n$. The options are summarized by

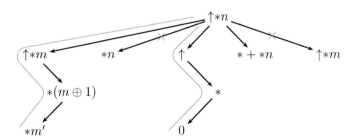

Since $n > 1$, Left has both $\uparrow$ and $\uparrow\!*$ as options. For any m, at least one of these will dominate $*m$. Note that, for $m < n$, $m \oplus 1 \neq n \oplus 1$, so by Lemma 9.2 $*(m \oplus 1) < \uparrow\!*n$. Hence, each Left option $\uparrow\!*m$ reverses to $*m'$, where each $m' \neq n \oplus 1$. Once bypassed, these options are all dominated by the option to $\uparrow$. Left's option to $\uparrow$ also reverses through $*$ to 0, and Left's option to $\uparrow\!*$ reverses out. This shows that the single Left option in the canonical form of $\uparrow\!*n$ is 0.

For Right, all options $\uparrow\!*m$ are dominated by $* + *n = *(n \oplus 1)$ by Corollary 9.3. Once the reader confirms that there are no further reversible options, we are done. $\square$

Example 9.6. In FORKLIFT, a single heap has value 0 since neither player has a move. If there are two heaps, $(1, a)$, then both Left and Right can only move to the position with one heap of size $a + 1$. Therefore, $(1, a)$ has value $*$. The position $(2, 2 + i)$ for $i \geq 0$ has options $\{*, 0 \mid *, 0\} = *2$.

Exercise 9.7. In FORKLIFT, show that $(a, a+i)$, for $i \geq 0$, has value $*a$.

The values of three heap positions are more complicated to describe. The position (p, q, p) for $p, q > 0$ has value 0 by the Tweedledum-Tweedledee strategy.

Exercise 9.8. Show that $(a, a+i, 1) = \uparrow\!*(a \oplus 1)$ for $a \geq 2$ and $i \geq 1$.

Exercise 9.9. Show that $(a+i, a, 1) = \uparrow\!*((a+1) \oplus 1))$ for $a \geq 2$ and $i \geq 1$.

G^n and sums of G^n

Definition 9.10. Let $G = \{0 \mid \mathcal{G}^R\}$ be a positive infinitesimal in canonical form. For a positive integer n define G^n by[1]

$$G^n \stackrel{\text{def}}{=} \{0 \mid \mathcal{G}^R - G^1 - G^2 - \cdots - G^{n-1}\}.$$

We will soon see that this defines a sequence of positive games, each infinitesimal with respect to the previous one; that is, for all positive integers m, $m \cdot G^{n+1} < G^n$.

Exercise 9.11. Find the canonical form for $\uparrow^2$.

For a fixed G, John Conway and Alex Ryba suggest the *uptimal notation:*[2]

$$.i_1 i_2 i_3 \ldots = i_1 G + i_2 G^2 + i_3 G^3 + \cdots.$$

For example, for the most common case when $G = \uparrow$, we have

$$.2013 = \Uparrow + \uparrow^3 + \uparrow^4 + \uparrow^4 + \uparrow^4.$$

For negative coefficients, place a bar over the integer:

$$.2\bar{1}1\bar{2} = \Uparrow - \uparrow^2 + \uparrow^3 - \uparrow^4 - \uparrow^4.$$

---

[1] The definition generalizes nicely if you use the *ordinal sum*, $G\!:\!H$, from the next chapter (Section 10.5 on page 239) and from [BCG01, p. 219]:

$$G\!:\!H \stackrel{\text{def}}{=} \{\mathcal{G}^L, G\!:\!\mathcal{H}^L \mid \mathcal{G}^R, G\!:\!\mathcal{H}^R\},$$
$$G^x \stackrel{\text{def}}{=} \{0 \mid G^R\!:\!(1-x)\},$$

for any number $x \geq 1$. When $x = n$ is a positive integer, the two definitions are equivalent. (Be warned that the definition of $G\!:\!H$ depends on the form of G, so be sure to use canonical form in G^x.)

[2] For generalized G^x, Conway and Ryba propose *uppity notation*, which uses raised digits as on a ruler to indicate fractional power terms. For instance,

$$.1^{2^3}0^{4^0} = .1^{2^3}0^0 4^0 0 = G^1 + 2 \cdot G^{1+1/2} + 3 \cdot G^{1+1/2+1/4} + 4 \cdot G^{2+1/4}.$$

Two such values are easily compared by inspecting the leftmost ruler mark on which they differ.

If the base G, say $G = {+}_2$, is not clear from context, you can specify it paren-thetically:

$$.i_1 i_2 i_3 \ldots \quad (\text{base} \quad _2).$$

Another notation, where

$$G^{\to n} = \underbrace{.11 \ldots 1}_{n} = G + G^2 + \cdots G^n,$$

is often convenient and appears in CGSuite. Since the time when *ONAG* [Con01] and *WW* [BCG01] were written, many more all-small games have been characterized and found to be important: Conway's uptimal notation is quite intuitive and is more general (see, for example, Problem 14). Not every all-small game corresponds to an uptimal expression though (see Problem 15).

Theorem 9.12. *Let base $G = \left\{ 0 \mid \mathcal{G}^R \right\}$ be a positive infinitesimal. Then, for all positive integers n and all non-negative integers m, $G^n > m \cdot G^{n+1}$.*

Proof: We will first prove that $G^n > 0$, by demonstrating that Left wins moving first or second from G^n. When $n = 1$, $G^1 > 0$ as asserted by the theorem. When $n > 1$, as the first player she, perforce, moves to 0, which fortunately wins. As the second player, she can reply to Right's move to some $G^R - G^1 - \cdots - G^{n-1}$ by playing on the G^{n-1}, leaving a total that is 0:

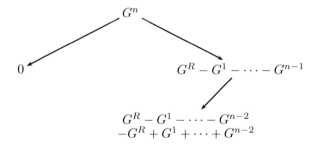

We will now show that $G^n - m \cdot G^{n+1} > 0$ by giving a winning strategy for Left as either the first or second player. As the first player she selects some option $G^R \in \mathcal{G}^R$ such that $G^R \lhd G$. One such G^R must exist, for otherwise G was a number by Theorem 6.15 on page 147. She then moves in one of the $-G^{n+1}$ components to $-G^R + G^1 + G^2 + \cdots + G^n$, leaving Right to move in

$$(G - G^R) + G^2 + G^3 + \cdots + G^{n-1} + 2 \cdot G^n = (G - G^R) + X.$$

We will now dispense with Right's possible responses. Suppose first that he makes a move in $G - G^R$. This is either to 0, to some $G^{R'} - G^R$, or to some $G - G^{RL}$. By induction, the first of these leaves a positive remainder, which Left wins easily. The second leaves a first-player win (since all the right options of G must be incomparable with one another) plus a positive remainder, and Left can win as first player. For the final case $G - G^{RL}$ cannot be negative, lest G^R be reversible, so again Left wins the position as first player.

If Right moves in some G^k that is part of X to $G^{R'} - G^1 - G^2 - \cdots - G^{k-1}$, then Left will be faced with $G^{R'} - G^R$ plus a positive remainder — and even if $R' = R$ will be able to win.

As second player, after Right's move, Left is faced either with

$$G^R - G^1 - G^2 - \cdots - G^{n-1} - m \cdot G^{n+1}$$

or with

$$G^n - (m-1) \cdot G^{n+1}.$$

The second case she wins by induction. In the first case Left can move in $-G^{n+1}$ to

$$-G^R + G^1 + G^2 + \cdots + G^{n-1} + G^n$$

leaving

$$G^n - (m-1) \cdot G^{n+1},$$

which again is an adequate reply by induction. □

Corollary 9.13. *The sign of any game represented in uptimal notation is the same as the sign of the first non-zero digit.*

Exercise 9.14. Prove Corollary 9.13.

Theorem 9.15. *Fix $n > 0$ and let base $G = \{0 \mid G^R\}$ be a positive infinitesimal in canonical form. The canonical form of $.\underbrace{11\ldots1}_{n}$ is given by*

$$.\underbrace{11\ldots1}_{n} = \left\{ .\underbrace{11\ldots1}_{n-1} \,\middle|\, \mathcal{G}^R \right\}.$$

Proof: See Problem 9.   □

In Section 5.4, we found that on the infinitesimal scale marked by ↑, the confusion interval of $*$ is between about ↓ and ↑. In particular, $* \parallel ↑$ but $* < ⇑$. We can now pin that confusion interval down more exactly. The following theorem tells us that when working base ↑,

$$.1111111 \quad *, \text{ but}$$

$$.1111112 > *;$$

so, on the infinitesimal scale, the confusion interval of $*$ is about

$$(.\overline{111}\ldots, .111\ldots).$$

Theorem 9.16. *Fix an infinitesimal canonical base* $G = \{0 \mid \mathcal{G}^R\} > 0$, *and fix* $G^R \in \mathcal{G}^R$. *Then,*

$$.111\ldots 11 \lhd G^R, \quad but$$

$$.111\ldots 12 > G^R.$$

More precisely, if $G \parallel G^R$, *then* $.111\ldots 11 \quad G^R$.

For example, $\uparrow + \uparrow^2 + \uparrow^3$ is incomparable with $*$, but $\uparrow + \uparrow^2 + \uparrow^3 + \uparrow^3 > *$.

Proof: Right moving first from $.111\ldots 11 - G^R$ can move to 0 by moving on the $.000\ldots 01$ and so win. But Right's dominant moves from $.111\ldots 12 - G^R$ are to $.000\ldots 01 + (G^{R'} - G^R)$ from which Left wins. What remains is the sum of a positive game $.000\ldots 01$ and a game $G^{R'} - G^R$ that, since G is in canonical form and has no dominated options, is either 0 or incomparable with 0.

When $G \quad G^R$, Left moving first rewrites $.111\ldots 11 - G^R = .011\ldots 11 + (G - G^R)$ and wins by some move on $G - G^R$ since $.011\ldots 11 > 0$. □

Corollary 9.17. *Working base* $\uparrow$, *fix* $S = .i_1 i_2 i_3 \ldots i_n$, *where every* $i_k \geq 1$ *and* $i_n > 1$. *Then, the canonical form of* $S + *$ *is given by* $\{0 \mid .j_1 j_2 j_3 \ldots j_n\}$, *where each* $j_k = i_k - 1$.

For example, $.11214* = \{0 \mid .00103\}$. As usual, we write $.11214*$ to mean $.11214 + *$.

Proof: See Problem 10. □

Exercise 9.18. In base $\uparrow$, show that $\underbrace{.11\ldots 1}_{n+1} * = \left\{ 0 , \underbrace{.11\ldots 1}_{n} * \,\middle|\, 0 \right\}$.

Example 9.19. We are now prepared to give a complete analysis of CUTTHROAT STARS. In particular, for $j > 0$ we have the following fact:

Theorem 9.20.

$$\multimap_j^i = \begin{cases} .\underbrace{jjj\ldots j}_{i} + i \cdot * & \text{if } i > 0, \\[2ex] .j - .1 + j \cdot * = (j-1) \cdot \uparrow* + * & \text{if } i = 0. \end{cases}$$

Proof: See Problem 13. □

For example,

$$\multimap_2^3 + \multimap_6^4 + \multimap_3^2 + \multimap_5^5 = .222* + .6666* + .\overline{33}* + .\overline{55555}* = .0031\overline{5},$$

which is a win for Left since the sign is positive.

All-small clear the pond

In ALL-SMALL CLEAR THE POND, one greedy strategy that might be worth trying is to always move a piece that goes farthest. However, this is not the optimal strategy. In ⬜●● ⬜●●●, Left has three options,

$$b = \boxed{\square \bullet\bullet\bullet\bullet\bullet}, \quad c = \boxed{\bullet\bullet \square \bullet\bullet}, \quad \text{and} \quad d = \boxed{\bullet\bullet \square \bullet \square \bullet}$$

where $b > d > c$, so the longest jump is not necessarily the best move.

While the game in general appears tough, we can find all the values for a game of ALL-SMALL CLEAR THE POND with only two pieces. After a few minutes thought, it becomes apparent that there seem to be several local situations on the board that affect the value. Clearly, jumping off the board is a move to zero. If the two pieces are adjacent and can jump, this seems to be hot (well, hot on the $\uparrow$ scale); if there are empty squares between the pieces, then the parity of the distance between them could be a factor.

Claim 9.21. $\square^m \bullet \square^n \bullet \square^{m+i} = i \cdot {\downarrow}{*} + *$ *by which we mean* $i \cdot ({\downarrow}{*}) + *$.

Proof: Freely using induction and the symmetric claim about $\square^{m+i} \bullet \square^n \bullet \square^m$,

$$\bullet \square^n \bullet = \{0 \mid 0\} = * = 0 \cdot {\downarrow}{*} + * ;$$

$$\square^m \bullet \square^n \bullet = \{0 \mid (m-1) \cdot {\uparrow}{*} + *\} = m \cdot {\uparrow}{*} + * ;$$

$$\square^m \bullet \square^n \bullet \square^{m+i} = \{(i-1) \cdot {\downarrow}{*} + * \mid (i+1) \cdot {\downarrow}{*} + *\}$$

$$= \{(i-1) \cdot {\downarrow}{*} + * \mid 0\}$$

$$= i \cdot {\downarrow}{*} + * .$$

In the second-to-last equality, Right's options reverse all the way to 0. $\square$

Claim 9.22.
$$\bullet\bullet\square = * ;$$
$$\bullet\bullet\square^i = (i-1) \cdot ({\downarrow}{*}) + * \qquad for\ i > 1.$$

Proof:
$$\bullet\bullet\square = \{\uparrow \mid 0\} = * ;$$
$$\bullet\bullet\square^i = \{(i-2) \cdot {\downarrow}{*} + * \mid 0\}$$
$$= (i-1) \cdot {\downarrow}{*} + * . \qquad \square$$

Claim 9.23. *For* $m > 0$ *and* $i \geq 0$,

$$\square^m \bullet\bullet \square^{m+i} = \{(i-2) \cdot {\downarrow}{*} + * \mid (i+2) \cdot {\downarrow}{*} + *\} .$$

Equivalently, when $m, n > 0$,

$$\square^m \bullet\bullet \square^n = \{{\Uparrow} \mid {\Downarrow}\} + (m-n) \cdot {\uparrow}{*} + * .$$

Proof: Writing out the options, we have

$$\square^m \bullet\bullet\bigcirc\square^{m+i} = \{\square^{m+1}\bullet\bullet\bigcirc\square^{m+i-1} \mid \square^{m-1}\bullet\bullet\bigcirc\square^{m+i+1}\}$$
$$= \{(i-2)\cdot{\Downarrow}*+* \mid (i+2)\cdot{\Downarrow}*+*\}. \qquad \square$$

9.2 Motivation: The Scale of Ups

The canonical forms of infinitesimal games that appear in practice are frequently quite complicated. In this sense, many of the infinitesimal positions that we have seen so far are, in fact, anomalous in their simplicity. However, it is often the case that an infinitesimal game consists of a sum of a simple dominant (or biggest) infinitesimal and a more complicated error term. For example, from Theorem 9.20 we know that

$$\circ\!\!\!-\!\!\!\overset{\bullet}{\underset{j}{\bullet}}^1 < \circ\!\!\!-\!\!\!\overset{\bullet}{\underset{j}{\bullet}}^2 < \circ\!\!\!-\!\!\!\overset{\bullet}{\underset{j}{\bullet}}^3 < \cdots$$

form an increasing sequence, but all are very nearly $j \cdot \uparrow$. If we know the dominant part of a number of infinitesimal games, then this is frequently sufficient information to determine the outcome of their sum.

Recall from page 80 that we defined $G = H$ if, for all games X, $G + X$ has the same outcome as $H + X$. When games are particularly complicated, it is often helpful to relax this definition. We could, for instance, define an equivalence relation in which two games are equivalent if they differ by only an infinitesimal.

Exercise 9.24. Review the definition of an equivalence relation. Define a relation $G \sim_{\text{inf}} H$ if $G - H$ is an infinitesimal, and quickly confirm that $\sim_{\text{inf}}$ is an equivalence relation.

If we are interested in understanding all-small games (or other infinitesimals), then this equivalence is too coarse, for all infinitesimals are then equivalent to 0. In the next section, we will define an equivalence that captures a great deal of what is important about all-small games. This equivalence relation will help us to describe any all-small game (and many infinitesimals that are not all-small) according to about how many $\uparrow$s the games are worth. We foreshadowed this when we showed a number line marked by multiples of $\uparrow$ in Figure 5.1 on page 120 and when we proved that $n\cdot\uparrow$ is larger than all infinitesimals born by day $n - 3$ (an easy consequence of Theorem 6.6 on page 143). In short, multiples of the game $\uparrow$ are a natural scale with which to measure infinitesimals.

9.3 Equivalence Under ✩

Somewhat surprisingly, it turns out that the equivalence relation we seek on infinitesimal games, which is intended to identify their dominant infinitesimal part, is most easily defined in terms of games played in the presence of a large NIM heap of indeterminate size. To prepare for this surprising definition, we first require the following theorem.

Theorem 9.25. *If $*m$ and $*n$ are not equal to any position of G (including G itself), then $G + *m$ has the same outcome as $G + *n$.*

Proof: It suffices to show that if Left wins moving first on $G + *m$, then Left wins moving first on $G + *n$.

   If Left wins moving first on $G + *m$ by moving to $G^L + *m$, Left also wins on $G^L + *n$ by induction. Suppose that Left wins by moving $G + *m$ to $G + *p$ for some $p < m$. If $*p$ is a position of G, then necessarily $p < n$ since $*n$ is *not* a position of G. So, in this case, the move to $G + *p$ is also available from $G + *n$. If, on the other hand, $*p$ is not a position of G, then since $G + *p \geq 0$ we also have

$$G = G + *p + *p \geq 0 + *p = *p.$$

But, $G \neq *p$ and so $G > *p$ is strict. Hence, Left wins moving first on $G + *p$, and thus also on $G + *n$, which has the same outcome as $G + *p$ by induction.   □

   Note that, in particular, the theorem applies if both m and n are larger than the birthday of G. This theorem justifies the following definition:

Definition 9.26. Two games are *far star equivalent*, $G \sim_✩ H$, if for all games X both $G + X + *n$ and $H + X + *n$ have the same outcome for n sufficiently large.

Again, by "sufficiently large" we mean that $*n$ is not the value of any position of $G + X$ or $H + X$. It suffices, for example, to select n to be greater than the birthday of both G and H.

   Repeating the phrase "$*n$ for n sufficiently large" is a bit of a nuisance. It becomes even more of a nuisance when adding two games each of which has a "$*n$ for n sufficiently large." It is therefore convenient (and natural) to introduce a symbol, ✩ (pronounced "far star"), to mean "$*n$ for sufficiently large n" and describe the algebraic properties of ✩. For example, when we add $G + ✩ + ✩$, each ✩ could, in turn, be set to be $*n_1$ and $*n_2$, respectively, from left to right, so that n_1 is sufficiently large in the context of G and n_2 is sufficiently large in the context of $G + *n_1$. But then, $*n_1 + *n_2$ can be replaced by a sufficiently large $*n$ without affecting the outcome, and so $G + ✩ + ✩$ should be the same as $G + ✩$, i.e., $✩ + ✩ = ✩$.

Let us pause right now to remark that ✦ *is not a short game.* In fact, it is not really a game at all. It serves very much the same notational purpose as the symbol ∞ serves in certain arguments about real numbers. We will use it in certain contexts as if it were a game, but the rules for working with ✦ will reflect its intended meaning — it stands for a NIM heap that is *large enough*, where the precise meaning of "large enough" depends on the surrounding context.

In mathematics, any quantity with the property that $a + a = a$ is termed an *idempotent*. For instance, considering the real numbers, under addition 0 is an idempotent, while under multiplication both 0 and 1 are idempotents. We have just argued that ✦ should be an idempotent.

In summary:

Definition 9.27. The symbol ✦ (*far star*) is defined to have the following properties in additive expressions involving ✦ and games:

- ✦ is an idempotent: ✦ + ✦ = ✦ .

- From ✦, a player may move to $*n$ for any value of $n \geq 0$.

The second property reflects the fact that playing $G + $✦ either player should, at least, be able to move to $G + *n$ whenever $*n$ is a position of G. The ability to move to even larger nimbers has no effect on the outcome of $G + $✦ by Theorem 9.25.

We now restate Definition 9.26.

Definition 9.28. For two games G and H, $G \sim_✦ H$ if $G + X + $✦ and $H + X + $✦ have the same outcome for all games X.

So, two games are equivalent in this sense if ✦ blurs them enough so that they behave the same in any context X.

Theorem 9.29. *The relation $\sim_✦$ is an equivalence relation that respects addition.*

Proof: The fact that $\sim_✦$ is reflexive, symmetric, and transitive is clear from the symmetries in the definition. To show that it respects addition, we need to show that if $G_1 \sim_✦ H_1$ and $G_2 \sim_✦ H_2$ then $G_1 + G_2 \sim_✦ H_1 + H_2$. For the latter, using $G_1 \sim_✦ H_1$, $G_1 + G_2 + X + $✦ has the same outcomes as $H_1 + G_2 + X + $✦ and, using $G_2 \sim_✦ H_2$, has the same outcomes as $H_1 + H_2 + X + $✦. $\square$

Neither definition of $\sim_✦$ is constructive. It would take a long time, indeed, to play G and H in every possible context $X + $✦. The following theorem comes to the rescue and also begins to suggest that $\sim_✦$ will provide the $\uparrow$ scale that we seek.

Theorem 9.30. $G \sim_✦ H$ *if and only if* $\downarrow < G - H < \uparrow$.

We will use one trick in the proof that is sufficiently subtle to warrant stating clearly in advance. According to Theorem 9.25, when playing a game of the form $J + \star$, Left can, at any time, insist that a $\ast$ be converted to any $\ast n$ as long as n is sufficiently large. This change will not affect the outcome of the game. In particular, Left can choose any n exceeding the birthday of J. This change can be safely made at any time and does not use up a turn. We will call this process *fixing the* $\ast$.[3]

Proof: For the $\Rightarrow$ direction, suppose that $G \sim_\star H$. Choosing $X = -G + \Downarrow$ in the definition of $\sim_\star$, we have that $G + X + \ast = \Downarrow \ast < 0$ has the same outcome as $H + X + \ast = G - H + \Downarrow \ast$. Hence, $(G - H) + \Downarrow \ast < 0$ and $G - H < \uparrow \ast$. A symmetric argument shows that $\downarrow \ast < G - H$.

The $\Leftarrow$ direction is more difficult. Suppose by way of contradiction that $\downarrow \ast < G - H < \uparrow \ast$ and yet it is *not* the case that $G \sim_\star H$. From the latter we may assume that, for some game X, Left wins $G + X + \ast$ moving first but cannot win $H + X + \ast$ moving first, and so $H + X + \ast \leq 0$. We will subscript the Xs to help track the games as they change: define $X_1 = X_2 = X$. We now have that

1. Right wins $-(G + X_1) + \ast$ moving first;

2. Right wins $(G + X_1) - (H + X_2) + \downarrow + \ast$ moving second;

3. Right wins $(H + X_2) + \ast$ moving second.

(The second item is a consequence of $G - H < \uparrow \ast$ and $X_1 - X_2 = 0$.)

Suppose now that Left plays all three of these games against three gurus. Left moves first in the second and third games but moves second in the first game. If Left wins any of these three games, we reach a contradiction.

Left's strategy is to let the gurus dictate play. At each point in the play, there will always be one guru whose move it is, and Left waits for that guru to make a move. If the guru cannot move, Left has won that game and we have established our contradiction. Based on the guru's move, Left then elects to move in one of the three games and waits for the response of the opposing guru on that board.

If a guru moves on $(G + X_1)$ in the second game, then Left can copy the move in $-(G + X_1)$ and vice versa. Similarly, moves on $(H + X_2)$ are matched to moves in $-(H + X_2)$. For the more challenging cases, the guru could move on a $\ast$ or move $\downarrow$ to 0. In response, respectively, Left will either move the remaining $\downarrow$ to $\ast$ or move on a $\ast$ (to anything). As described in the paragraph

---

[3]This is really a version of the One-Hand-Tied Principle. In fixing the $\ast$, Left agrees not to use some of the options available to her from $\ast$ and, moreover, promises that if Right makes a move in $\ast$ to $\ast m$ for some $m > n$, then she will respond immediately to $\ast n$.

preceding this proof, Left will now *fix* the remaining two ☆s. In particular, Left can do so in such a way that the three positions are now in the forms

$$-G' - X_1' + *n_1, \qquad (G' + X_1') - (H' + X_2') + *n_2, \qquad H' + X_2' + *n_3,$$

where Left ensures that $*n_1 + *n_2 + *n_3 = 0$. In particular, since a guru chooses at most one of the n_i, Left is free to fix the remaining two to be "sufficiently large" while maintaining that the sum $*n_1 + *n_2 + *n_3$ is 0.

After the ☆s are fixed in this way, the sum of the resulting games is 0, so Left has no trouble finding a natural response to any opposing move on an outer board by moving on the symmetric term in the inner board, and vice versa. Throughout play, Left can always find a move in one of the three components after a guru has moved, and so she never runs out of moves until after one of the gurus has run out and lost. □

9.4 Atomic Weight

Definition 9.31. If $g \sim_{\hat{\star}} G \cdot \uparrow$ then we say that g has *atomic weight* G and write $G = \mathrm{AW}(g)$.

In this section, we will use an uppercase letter to represent the atomic weight of a game written with a lowercase letter, since g is infinitesimal while G typically is not.

To implement this definition, we can *guess* the atomic weight G of g, compute $G \cdot \uparrow$, and use Theorem 9.30 to test whether $g \sim G \cdot \uparrow$. When using the unit $U = \uparrow$, the definition of Norton product requires U's incentives. The left incentive from $\uparrow$ is $\downarrow$, and the right incentive is $\uparrow*$. The latter is dominant, and we have $\tau = \{\Uparrow*\}$. So, the Norton product reduces to

$$G \cdot \uparrow = \begin{cases} \overbrace{\uparrow + \uparrow + \uparrow + \cdots + \uparrow}^{n} & \text{if } G = n \text{ is a positive integer,} \\ 0 & \text{if } G = 0, \\ \underbrace{\downarrow + \downarrow + \downarrow + \cdots + \downarrow}_{n} & \text{if } G = -n \text{ is a negative integer,} \\ \{G^L \cdot \uparrow + \Uparrow* \mid G^R \cdot \uparrow + \Downarrow*\} & \text{otherwise.} \end{cases}$$

Theorem 9.32. *The atomic weight G of g is well defined. That is, if G_1 and G_2 are both atomic weights of g, then $G_1 = G_2$.*

Proof: We will show that if $G_1 \cdot \uparrow \sim G_2 \cdot \uparrow$, then $G_1 = G_2$. Since the Norton product is additive, it suffices to show that if $G \cdot \uparrow \sim 0$ for G in canonical form, then $G = 0$.

Assume, by way of contradiction, that $G \neq 0$; then one player, say Left, wins moving first on G. This yields a winning strategy on $G \cdot \uparrow + \downarrow\star$: Left pretends that she is playing G alone and plays the corresponding moves in $G \cdot \uparrow$ until it becomes an integer multiple of $\uparrow$.

In essence, using this strategy, players benefit by the amount of $t = \Uparrow\star$ when playing on the Norton product (that is, Left gains t while Right gains $-t$), while Right can benefit by at most $\uparrow\star$ by playing on $\downarrow$ (or less by playing on   ). With careful accounting (see below), one can then confirm that once G reaches an integer, either Right is to move from a game of value > 0 or Left is to move from a game of value $\rhd\, 0$. Either way, Left wins.

Now, for the accounting. The key is that each move on $G \cdot \uparrow$ picks up $\Uparrow\star$ from the definition of Norton product.

- If Right always responds locally on $G\cdot\uparrow$, then since Left wins on G, when G first becomes an integer, say n, it is with Left having moved last and $n \geq 0$ or with Right having moved last and $n \geq 1$. So, Left achieves a game $\geq \Uparrow\star + \downarrow \;=\; \uparrow$ moving second or a game $\geq \uparrow + \downarrow \;=\;$ moving first. Either way she wins.

- If Right ever moves on $\downarrow$, Left gets two moves in a row in the $G \cdot \uparrow$ component. By Corollary 5.38, the integer left stop minus the integer right stop from G is at most -1 (and is only -1 when Left moves last from the left integer stop). Hence, each time Left plays twice in a row in the Norton product, she loses no more than $\downarrow$ from the stops but picks up $\Uparrow\star$ from the Norton product definition. Right, however, picks up at most $\downarrow\star$ for his move in the second component. In total, Left does no worse than the analysis in the preceding paragraph. □

Theorem 9.33. *Atomic weights are additive:* $\mathrm{AW}(g) + \mathrm{AW}(h) = \mathrm{AW}(g + h)$.

Proof: This is an immediate consequence of the fact that $\sim_\star$ respects addition. □

Again, we can use Theorem 9.30 along with the definition of atomic weights to confirm that a conjectured atomic weight is correct.

Example 9.34.

- It is trivially the case that $\mathrm{AW}(G \cdot \uparrow) = G$.
- $\mathrm{AW}(\star) = 0$ for $\downarrow$ $< \star < \uparrow$.
- $\mathrm{AW}(\uparrow\star) = \mathrm{AW}(\uparrow) + \mathrm{AW}(\star) = 1 + 0 = 1$.
- $\mathrm{AW}(\pm\Uparrow) = \mathrm{AW}(\pm\Uparrow\star) + \mathrm{AW}(\star) = \star + 0 = \star$, because $\star \cdot \uparrow = \pm\Uparrow\star$.

Exercise 9.35. Determine AW($\pmb{+}_2$) and AW($\{0 \mid \quad_2\}$). (Both are in $\{-1, 0, 1\}$.)

Exercise 9.36. Show that AW($.pq$) $= p$, where $.pq$ is in uptimal (base $\uparrow$).

It is important to note that not all infinitesimal games have an atomic weight.

Example 9.37. If $g = \{1 \mid \Uparrow \,\|\, \Downarrow \mid -1\}$, there is no G such that $G = \text{AW}(g)$; for example, suppose that there were a G such that

$$\downarrow\!\!\star \,<\, g - G \cdot \uparrow \,<\, \uparrow \ .$$

When Left wins moving second from $g - G \cdot \uparrow - \downarrow$, Left must reply to a move on g (for -1 is less than any option of the all-small game $G \cdot \uparrow$). This leaves $\Downarrow - G \cdot \uparrow + \uparrow$ from which Left should win moving second. So, $\downarrow \ \geq G \cdot \uparrow$. A symmetric argument for Right yields

$$\downarrow \ \geq G \cdot \uparrow \geq \uparrow \ ,$$

which is impossible.

Knowing the atomic weight of a game does not necessarily determine the outcome. For example, all four games $g \in \{0, *, \uparrow^2, -\uparrow^2\}$ have atomic weight 0 but have different outcomes. However, a sufficient advantage in atomic weight is decisive.

Theorem 9.38. (The Two-Ahead Rule)

$$
\begin{array}{llll}
\textit{If} & \text{AW}(g) \geq +2 & \textit{then} & g > 0. \\
\textit{If} & \text{AW}(g) \geq +1 & \textit{then} & g \,\triangleright\, 0. \\
\textit{If} & \text{AW}(g) \leq -1 & \textit{then} & g \,\triangleleft\, 0. \\
\textit{If} & \text{AW}(g) \leq -2 & \textit{then} & g < 0.
\end{array}
$$

Proof: If $G = \text{AW}(g) \geq 2$, then $g - \Uparrow \geq g - G \cdot \uparrow > \downarrow$ using Theorem 9.30 for the second inequality. This gives $g > \uparrow \ \geq 0$. When $\text{AW}(g) \geq 1$, the same line of reasoning yields $g > \quad 0$. The third and fourth assertions are symmetric. □

While Theorem 9.30 gives a way to test whether g has atomic weight G, it gives no guidance for how to find this atomic weight G, or even whether it exists. Example 9.37 gave an infinitesimal with no atomic weight, but every all-small game does, in fact, have an atomic weight and [BCG01] proves that the following mechanical method computes it!

Theorem 9.39. *If $g = \{g^L \mid g^R\}$ and g^L and g^R have atomic weights, then the atomic weight of g is given by*

$$\left\{ \mathrm{AW}(g^L) - 2 \mid \mathrm{AW}(g^R) + 2 \right\}$$

unless this is an integer. In that case, let x be the least integer such that $\mathrm{AW}(g^L) - 2 \triangleleft\!| \ x$ and y the greatest integer such that $y \quad \mathrm{AW}(g^R) + 2$.

- *If $g \parallel \bigstar$, $\mathrm{AW}(g) = 0$.*

- *If $g > \quad$, $\mathrm{AW}(g) = y$.*

- *If $g < \quad$, $\mathrm{AW}(g) = x$.*

In practice, atomic weights are often integers, and the exception is invoked.

Proof: We refer you to the proof in [BCG01, pp. 248–251]. If you work through their proof, be aware that they use the symbol $\doteq$ where we use $\sim_{\bigstar}$ and that their $\stackrel{\wedge}{\bigstar}$ is our   . □

To give credibility to this theorem, let's try a few examples:

- AW(0): $\{-2 \mid 2\} = 0$ and since $0 \qquad$ then $\mathrm{AW}(0) = 0$.

- AW($*$): $\{0 - 2 \mid 0 + 2\} = 0$ and since $* \qquad$ then $\mathrm{AW}(*) = 0$. By identical reasoning (now employing induction), $\mathrm{AW}(*n) = 0$.

- AW($\uparrow$): $\{0 - 2 \mid 0 + 2\} = 0$ but $\uparrow > \quad$. So, $\mathrm{AW}(\uparrow) = 1$, the largest integer less than or incomparable with 2.

- AW($\uparrow\!*n$): $\{0 - 2 \mid 0 + 2, 1 + 2\} = 0$ but since $\uparrow\!*n > \quad$, $\mathrm{AW}(\uparrow\!*) = 1$, the largest number less than or incomparable with both 2 and 3.

- AW($\Uparrow$): $\Uparrow = \{0 \mid \uparrow\!*\}$, so the purported atomic weight is $\{0 - 2 \mid 1 + 2\} = 0$, and since $\Uparrow > \quad$ we have $\mathrm{AW}(\Uparrow) = 2$.

- AW(.11): Working base $\uparrow$, $\mathrm{AW}(.11) = \mathrm{AW}(\{\uparrow \mid *\})$. So, the purported atomic weight $\{1 - 2 \mid 0 + 2\} = 0$ again, and since $.11 > \uparrow > \quad$ we have $\mathrm{AW}(.11) = 1$.

Exercise 9.40. Use Theorem 9.39 to determine $\mathrm{AW}(\{(n+1)\cdot\uparrow \mid n\cdot\uparrow\})$ for all integers n. In other words, compute the atomic weights of

$$\ldots, \{\downarrow \mid \Downarrow\}, \{0 \mid \downarrow\}, \{\uparrow \mid 0\}, \{\Uparrow \mid \uparrow\}, \{\Uparrow\!\!\!\uparrow \mid \Uparrow\}, \ldots.$$

Exercise 9.41. Let

$$g = \{0, *, *2, \ldots, *m \mid 0, *, *2, \ldots, *n\}.$$

Use Theorem 9.39 to show that if $m > n$ then $\mathrm{AW}(g) = 1$.

ALL-SMALL CLEAR THE POND had some hot values (on the infinitesimal scale). The atomic weights of these games are interesting, and we show how to interpret (use) them later in Lemmas 9.47 and 9.52. In the following examples, the exceptional case of Theorem 9.39 fails to occur, and so

$$\mathrm{AW}(\{\Uparrow* \mid \Downarrow*\}) = \{2 - 2 \mid -2 + 2\} = *;$$
$$\mathrm{AW}(\{\uparrow \mid \Downarrow\}) = \{1 - 2 \mid -3 + 2\} = -1*.$$

9.5 All-Small Shove

This variant of SHOVE has the same rules except that the game is deemed to be over when the last of either player's pieces disappears from the board. For example, in SHOVE, the position

has value $4\frac{1}{2}$, but in ALL-SMALL SHOVE

$$\boxed{\quad} = \left\{ \boxed{\quad} \;\middle|\; \boxed{\quad} \right\} = \{0 \mid 0\} = *.$$

Unlike SHOVE, in this game the rightmost piece does not determine the winner of a single component. For any position in which the rightmost squares contain at least one blue piece and no red pieces, the squares can be replaced by a single square with a single blue piece:

The game will end when the rightmost red piece leaves the board, and any move to the right of this piece has the same effect. So, without loss of generality, we need only consider positions in which the rightmost two pieces are adjacent and opposite in color.

As usual, let $\boxed{}^{n}$ denote a row of n empty squares.

Exercise 9.42. Show that in ALL-SMALL SHOVE

$$\boxed{}^{n}\boxed{}\boxed{} = (n+1) \cdot *.$$

Exercise 9.43. Show that in ALL-SMALL SHOVE

$$\boxed{}^{m}\boxed{}\boxed{}^{n}\boxed{}\boxed{} = (m+1) \cdot \uparrow + (n+m) \cdot *.$$

Exercise 9.44. Show that in ALL-SMALL SHOVE

$$\square^m\,\boxed{\leftarrow}\,\square^n\,\boxed{\leftarrow}\,\boxed{\leftarrow} = (m+1)\cdot\downarrow + (n+m)\cdot *.$$

While some positions in this game are easily described using uptimal notation, others appear quite complex for current techniques:

Position	Value	Atomic weight
(dominoes)	$\downarrow$	-1
(dominoes)	$\{0 \mid \uparrow, \Uparrow *\}$	2
(dominoes)	$.21*$	2
(dominoes)	$.321$	3
(dominoes)	$.4321*$	4
(dominoes)	$\uparrow$	1
(dominoes)	$.21*$	2
(dominoes)	$\{\downarrow, \Downarrow * \mid 0\}$	-2
(dominoes)	$0 \parallel 0 \mid .21*, \{0 \mid .21*\} \mid\mid\mid 0 \mid .21*, \{0 \mid .21*\}$	4

9.6 More Toppling Dominoes

Consider TOPPLING DOMINOES positions with blue, red, and green dominoes. (The green dominoes can be tipped by either player.)

Exercise 9.45. The outcome of a single component in TOPPLING DOMINOES is completely determined by the colors of the two ends. Show how.

We have the following astonishingly simple result:

Theorem 9.46. *Any* TOPPLING DOMINOES *component in which both ends are green has atomic weight 1, 0, or* -1.

What makes this theorem surprising is that the values of even small TOPPLING DOMINOES positions can have quite complicated canonical forms. For instance,

$$\text{(dominoes)} = \left\{ \begin{array}{c} 0,\ \{1\mid 0\},\ \{1*\mid 0\}, \\ \{1, 1*, \{1, 1* \mid 0, \{1\mid 0\}\} \mid 0, *, \{1, \{1\mid 0\} \mid 0, *\}\}, \\ \{0, \{1\mid 0\}, \{1, 1*\mid 0, \{1\mid 0\}\} \mid 0\} \end{array} \middle| 0, * \right\}.$$

Before we prove the theorem, we start with another lemma. In particular, were we to pretend that TOPPLING DOMINOES were all-small, we would be led to proving the following:

Lemma 9.47. *Any all-small game of the form*

$$\{0, \{? \mid 0\}, \ldots \mid 0, \{0 \mid ?\}, \ldots\}$$

has atomic weight 1, 0, *or* −1.

Note that the TOPPLING DOMINOES under consideration are of this form *except* that the question marks are not all-small.

Proof: The preliminary atomic weight calculation given by Theorem 9.39 yields

$$\text{``} \{-2, \text{`}\{? \mid 0\}\text{'}, \ldots \mid 2, \text{`}\{0 \mid ?\}\text{'}, \ldots\} \text{''},$$

where these question marks are the atomic weights of the previous question marks shifted by −4 and 4. Now the actual atomic weights of the inner quotation marks either have right option 0 or (in the exceptional case) are integers. Either way, the game in double quotes is a second-player win and has value 0. So, we are in the exceptional case. But, the only integers that can possibly fit between the left options and right options are 1, 0, and −1. ☐

Proof of Theorem 9.46: Let g be any green-ended TOPPLING DOMINOES position, and let g' be the same game with every position that is a positive number replaced by ⋔ and every negative integer replaced by ⫫. According to the lemma, the atomic weight of g' is 1, 0, or −1. Observe that $g - \Uparrow\!\!\star$ has the same outcome as $g' - \Uparrow$ and that the same statement can be made of ↑ , , ↓ , and ⫫ . Therefore, by Theorem 9.30, $g \sim_\star g'$ and they have the same atomic weight. ☐

Note that the proof gives no efficient way to determine which atomic weight is correct, 1, 0, or −1. The best method we know to date is to compare the game to   and find out.

9.7 Clobber

Generally, by about halfway through a game of CLOBBER, the board has decomposed into a disjunctive sum. Our techniques should be of great use when trying to play this game, but CLOBBER is a hard game! As we will see, even the atomic weights can be bad. We present some results and conjectures about CLOBBER, then we show how to find the atomic weight of a position that has exactly one red piece.

The one-dimensional version decomposes almost from the first move. Remember our usual convention: if X is a pattern of blue, red, and empty spaces, we denote a string consisting of n copies of this pattern by X^n, so

$$\left(\blacksquare\blacksquare^2\right)^3 = \blacksquare\blacksquare\blacksquare\blacksquare\blacksquare\blacksquare\blacksquare\blacksquare\blacksquare\blacksquare$$

Consider the one-dimensional version with just one red piece at the end of a string of blue pieces. The first few values are easy to calculate:

$$
\begin{aligned}
\blacksquare &= \blacksquare = 0; \\
\blacksquare\blacksquare &= *; \\
\blacksquare\blacksquare\blacksquare &= \{0 \mid \blacksquare\blacksquare\} = \{0 \mid *\} = {\uparrow}; \\
\blacksquare\blacksquare\blacksquare\blacksquare &= \{0 \mid \blacksquare\blacksquare\blacksquare\} = \{0 \mid {\uparrow}\} = {\Uparrow}*.
\end{aligned}
$$

We have seen this sequence before. In general:

Lemma 9.48.

$$\blacksquare\blacksquare^n = \left\{0 \mid \blacksquare\blacksquare^{n-1}\right\} = \{0 \mid (n-2)\cdot{\uparrow} + (n-1)\cdot*\} = (n-1)\cdot{\uparrow}+*.$$

Exercise 9.49. Show that if $m > 1$ and $n > 1$, then $\blacksquare^m\blacksquare^n = 0$.

Note, in particular, that $\blacksquare\blacksquare^n$ can be much bigger than $\blacksquare\blacksquare\blacksquare^n$, so a single stone can make a dramatic difference. Another natural sequence to investigate is alternating blue and red stones:

$$
\begin{aligned}
\blacksquare\blacksquare\blacksquare &= *; \\
\blacksquare\blacksquare\blacksquare\blacksquare &= \pm(*, {\uparrow}); \\
\blacksquare\blacksquare\blacksquare\blacksquare\blacksquare &= .\overline{1}\overline{1}; \\
\blacksquare\blacksquare\blacksquare\blacksquare\blacksquare\blacksquare &= 0; \\
\blacksquare\blacksquare\blacksquare\blacksquare\blacksquare\blacksquare\blacksquare &= \{{\downarrow}, \pm(*, {\uparrow}) \mid 0\}; \\
\blacksquare\blacksquare\blacksquare\blacksquare\blacksquare\blacksquare\blacksquare\blacksquare &= \pm(.11, \{{\uparrow}*, {\Uparrow} \mid 0, {\uparrow}*, \pm(0, {\uparrow}*)\}).
\end{aligned}
$$

The canonical forms get worse as the length of the board grows. Do the atomic weights help? Starting with $\blacksquare\blacksquare\blacksquare$, extending by one stone each time and continuing until there are 20 stones, the atomic weights are $0, 0, -1, 0, -1, 0, -1, 0, -1, 0, -1, 0, -1, 0, -1, 0$, and -1. Even though a one-dimensional board with a repeating pattern of pieces looks set up for induction, at the time of writing, none of the values, atomic weights, and outcomes of an alternating board are known.

The reader may wish to try proving either or both of the following:

Conjecture 9.50. $(\blacksquare\blacksquare)^n$ *is a first-player win for* $n \neq 3$.

Conjecture 9.51. ( )$^n = \lfloor (n+1)/2 \rfloor \cdot \uparrow$.

The first conjecture has been verified by computer up to $n = 19$, and, except for $n = 3$, the first player has few losing moves. The second conjecture has been verified up to $n = 17$.

For $n > 4$, the game

has atomic weight $\{0 \mid \{0 \mid 4-n\}\} = +_{n-4}$. In addition, positions with atomic weights of $\frac{1}{2}, \frac{1}{4}, \frac{1}{8}, \uparrow, *,$ and $_{1/2}$ can fit on a 6×3 board. The game

has atomic weight $\{n - 3 \mid 0\}$, showing that arbitrarily hot atomic weights can occur.

In the rest of this section, we will consider boards with only one red counter. We will also think of the board as a graph where squares on the board are vertices and adjacent squares are connected by edges. Consider a graph G and the corresponding game G_v in which all but vertex v is occupied by blue counters and v is occupied by a red counter.

A *blocked v-path* in G is any maximal path $P = \{v_0 = v, v_1, v_2, \ldots, v_r\}$ starting at v. That is, every vertex adjacent to v_r is in P so that the path cannot be extended from v_r. A *blocked split v-path*, or *split path* for short, is a blocked path $P = \{v_0 = v, v_1, v_2, \ldots, v_r\}$ for which there exist vertices x, y such that $Q = \{v_0 = v, v_1, v_2, \ldots, v_{r-1}, x, y\}$ is also a blocked path. The *length*, $l(P)$, of a blocked path $P = \{v_0 = v, v_1, v_2, \ldots, v_r\}$ depends on whether the path is split:

$$l(P) = \begin{cases} r+1 & \text{if } P \text{ is not a split path,} \\ r & \text{if } P \text{ is a split path.} \end{cases}$$

It may be that the two paths in a split path only have v in common, then $l(P) = 1$; for example, . Let

$$l(G_v) = \min\{l(P) \mid P \text{ is a blocked path or a split path of } G_v\}.$$

For example,

$$l(\text{}) = 3;$$
$$l(\text{}) = 4;$$
$$l(\text{}) = 5;$$
$$l(\text{}) = 6.$$

Note that if G is a path of $n \geq 2$ vertices with v at one end, then Lemma 9.48 gives that $\text{AW}(G) = \text{AW}((n-2)\cdot\uparrow + n\cdot*) = n - 2$. Also, note that $\mathrm{AW}(\boxed{\text{▨▨▨▨}}) = \mathrm{AW}(\downarrow *) = -1$, whereas $\mathrm{AW}(\boxed{\text{▨▨▨▨▨▨}}) = 1$. At the end of a split path (i.e., at $\boxed{\text{▨▨▨▨}}$), Right has the option of moving to 0 or to $\boxed{\text{▨▨}} = *$, both of which have atomic weight 0. This choice between two apparently small games is actually the difference of playing to a first-player-win and a second-player-win game. This gives Right a local advantage that is reflected in the atomic weights.

Lemma 9.52. *For a graph G with $v \in V(G)$ having at least one neighbor,*

$$\mathrm{AW}(G_v) = l(G_v) - 2.$$

Proof: We prove this by induction on the number of squares and assume throughout that G is connected and v has a neighbor. Note that Left's options are all identically 0.

First, suppose that $l(G_v) = 1$. In particular, G_v is of the form

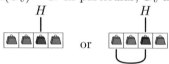

where H is arbitrary and possibly empty. Right has moves to $*$ and to 0. A move to a vertex in H by induction has atomic weight ≥ 0. In the preliminary atomic weight calculation of G_v (given by Theorem 9.39 on page 221),

$$G_0 = \{0 - 2 \mid 0 + 2, \ldots\},$$

and the atomic weight of G_v is either -1, 0, or 1 depending on how G_v compares with ☆. But,

$$G_v = \{0 \mid 0, *, \ldots\} \leq \{0 \mid 0, *\} = \downarrow *,$$

which has atomic weight -1. So, $\mathrm{AW}(G_v) = -1$.

Henceforth, we assume that $l(G_v) \geq 2$. We know that

$$G_v = \{0 \mid (G - v)_w, \ w \text{ adjacent to } v\}.$$

There is at least one vertex b that is the next vertex on a shortest blocked path from v, and $l((G-v)_b) = l(G_v) - 1$. So, the preliminary atomic weight of G_v is

$$G_0 = \{0 - 2 \mid l((G-v)_b - 2 + 2\} = \{-2 \mid l(G_v) - 1)\},$$

and $\mathrm{AW}(G_v)$ is -1, 0, or $l(G_v) - 2$, depending on how G_v compares with   . If $l(G_v) = 2$, then $\mathrm{AW}(G_v)$ is restricted to just -1 and 0. Since Left can win moving first on $G_v +$ by moving to G_v, $\mathrm{AW}(G_v) = 0 = l(G_v) - 2$. Left wins moving first or second when $l(G_v) > 2$ by removing   at the earliest opportunity. Hence, in this case, $\mathrm{AW}(G_v) = l(G_v) - 2$. $\qquad\square$

Problems

1. Find the value of

2. Find the value of

3. Find the values of the following ALL-SMALL PUSH positions. Express them using uptimal notation (base ↑) from page 210.

(a) [♦ | | ♦]

(b) [| ♦ | ♦]

(c) [♦ | ♦ | ♦]

4. Who wins the disjunctive sum of the ALL-SMALL CLEAR THE POND position

,

the CUTTHROAT STARS position

,

and the ALL-SMALL SHOVE position

[♦ | ♦ | ♦ | ♦] ?

5. In ALL-SMALL RUN OVER show that

$$\bullet\bullet^{m} = .\underbrace{\bar{1}\bar{1}\ldots\bar{1}}_{n}*,$$

where $n = \frac{m(m-1)}{2}$.

6. In ALL-SMALL DRAG OVER show that

$$\bullet^{m}\bullet^{n} = .\underbrace{mm\ldots m}_{n-1} + m\cdot* = m\cdot(.\underbrace{11\ldots1}_{n-1}*).$$

7. In the partizan version of FORKLIFT, find the values for all positions of the form $(1, 1, \ldots, 1)$.

8. In PARTIZAN ENDNIM find the value for the one-pile position a and the two-pile positions of the form $b1$.

9. Prove Theorem 9.15 on page 212.

10. Prove Corollary 9.17 on page 213. If your proof is overly confusing, you might want to also demonstrate your proof using .11214 as an example.

11. Let $X = .\underbrace{11 \ldots 1}0Y$ be a string of 0s and 1s, and let k be a positive integer. Express each of the following as an uptimal (base ↑), perhaps plus *:

    (a) $\{0, .X* \mid 0\}$;

    (b) $\{0, .X\bar{k}* \mid 0\}$.

    These are generalizations of Exercise 9.18.

12. Prove that a game in canonical form that is not all-small is not equal to any all-small game. (*Hint:* Prove that if G is all-small, then it remains all-small when converted to canonical form.)

13. Prove Theorem 9.20.

14. Consider CUTTHROAT played on a path. Show that

    (a) $RL^n = .\underbrace{11 \ldots 1}_{n-1}*$,

    (b) $LRL^n = .\underbrace{00 \ldots 0}_{n-1}\bar{1}.$

15. Show that $\{0 \mid *, *2\}$ cannot be expressed as a (base ↑) uptimal plus a nimber by showing that, for all non-negative j and k,

    (a) $\{0 \mid *, *2\} \triangleright .0k + *j$ and

    (b) $\{0 \mid *, *2\} \triangleleft .1\bar{k} + *j.$

16. Show that the values, in uptimal notation, of the ALL-SMALL BOXCARS positions are

    (a) ⬤◻⬤ = *,

    (b) ⬤◻⬤◻⬤ = .1,

(c) = .2*,

(d) = .31.

17. In the following disjunctive sum of an ALL-SMALL BOXCARS, a CLOBBER, and a CUTTHROAT STARS position, is there a winning move for Left? For Right?

Preparation for Chapter 10

If we know the canonical form of a game then we know, at least in principle, who should win. But computing a full canonical form can be a lot of work! Sometimes if we're only interested in a winner then shortcuts can be taken. In particular, if Left has a "large" advantage in some part of the game, then she may be able to ignore the rest of it.

Prep Problem 10.1. Often (or perhaps almost always), in order to analyze a complex system, we deliberately simplify the system, choosing to lose information in the process. Sometimes this is explicit, as in "assume frictionless motion." Sometimes this is implicit, "are you in pain?" Try to come up with examples of this strategy in a few different fields of human endeavours, some analytical, some not.

To the instructor: The two major concepts introduced in this chapter are *reduced canonical form* and *ordinal sum*. Both allow for the simplification of game trees in some cases, producing games with different values but somehow close to the original ones. Such approximations are important in the preliminary analysis of many games, in particular to determine the winner.

Chapter 10

Trimming Game Trees

Natural abilities are like natural plants; they
need pruning by study.

Red Auerbach

10.1   Introduction

In Section 3.3 we discussed the question of when two games should be considered equal, and we eventually settled on the viewpoint that for our purposes "equality" should mean the same thing as "having the same canonical form." But, the canonical form a game, G, captures *all* the relevant information about the game tree of G required to play G in isolation or as part of any sum. In many situations not all of this information is required. For instance, if the value of G is 3 plus an infinitesimal, then the players already know what the outcome should be if $G + x$ is played for any number $x \neq -3$. Furthermore, in many games there are incomparable options that only differ by complex infinitesimals, and these differences only affect strategy in esoteric or carefully constructed game positions.

One technique to uncover this structure is to ignore infinitesimal differences between options and make a first-order approximation by restricting our attention on the underlying numbers and hot games. In this chapter, we will introduce the *reduced canonical form* (rcf) of a game, that being the simplest value infinitesimally close to the game. In Section 10.3 we will be able to understand a great deal about MAZE positions by evaluating their reduced canonical forms and compromising a little on the completeness of the analysis.

Another situation that raises similar issues occurs in games where there is some notion of "small moves" (which change the structure of the game only a little) and "big moves" (which change it a lot). For example, in HACKENBUSH played on a tree, it is a small move to cut an edge that is far from the ground, while it's definitely a big move to cut the edge connecting the tree to the ground (since that ends the game immediately). This concept is captured in a new definition of game composition, the *ordinal sum*. Rather than adding two games as independent equals, one can staple the game tree of one game (the branch) onto the leaves of another game tree (the base). The base game exhibits the dominant structure, and the branch game adds substructure. We will show how this idea can be used to analyse the game of STIRLING-SHAVE (Section 10.6) and also to characterize the positions in TOPPLING DOMINOES whose values are numbers (Section 10.7).

10.2   Reduced Canonical Form

As we suggested in the previous section, the reduced canonical form of a game G should be the simplest game that differs from G by an infinitesimal. We will see later how to compute this value, denoted $\mathrm{rcf}(G)$, while simultaneously demonstrating that it is well defined, but first it seems natural to ask, *Why is the reduced canonical form useful?*

The main thing is that it ignores infinitesimals in the canonical form. This means $\mathrm{rcf}(G)$ is an approximation of the value of G. In particular, if $\mathrm{rcf}(G) > \mathrm{rcf}(H)$ then we know that $G > H$ and hence that Left wins $G - H$. The problem arises when $\mathrm{rcf}(G) = \mathrm{rcf}(H)$. In that case $G - H$ might be of any one of the four outcome classes, and so we gain no information about the relative values of G and H. However, $G - H$ will be an infinitesimal, so if a player loses it will only be because the opponent got the last move without any spare moves remaining. The game will be close! In this chapter we will give the broad strokes of the theory of reduced canonical form, and we leave the reader to consult Grossman and Siegel's work [GS07, Sie13] for further details.

Recall from Problem 3 of Chapter 6 that X is an *infinitesimal* if $LS(X) = RS(X) = 0$. Since we will use this fact frequently in this chapter to identify infinitesimals when comparing games, we restate it here using a game difference:

Theorem 10.1. $G - H$ *is infinitesimal if and only if* $LS(G - H) = RS(G - H) = 0$.

We need some way of referring to games without worrying about the infinitesimals.

Definition 10.2. When $G - H$ is infinitesimal, we say that G and H are *infinitesimally close*, and we write $G \sim_\epsilon H$. We can equivalently say that H is *G-ish* (*G* Infinitesimally SHifted). A game is *number-ish* if it is infinitesimally close to a number.

For inequalities, we write $G \geq_\epsilon H$ if $G \geq H + \epsilon$ for some infinitesimal ϵ; $G \leq_\epsilon H$ is defined similarly.

Since the sum and difference of infinitesimals are infinitesimal and 0 is infinitesimal, it is clear that $\sim_\epsilon$ is an equivalence relation and $\geq_\epsilon$ is a partial order.

Exercise 10.3. Confirm the preceding claims.

Corollary 10.4. *A game G is number-ish if $LS(G) = RS(G)$.*

Proof: Suppose that $LS(G) = RS(G) = x$ and let $H = G - x$. By the Number-Translation Theorem (Theorem 6.13 on page 6.13), $LS(H) = RS(H) = 0$, so H is an infinitesimal and therefore G is number-ish.   □

Theorem 10.5. *Let G and H be games. If $\mathbf{RS}(G) \geq \mathbf{LS}(H)$ then $G - H \geq \epsilon$ for some infinitesimal ϵ; that is, $G \geq_\epsilon H$.*

Proof: If $\mathbf{RS}(G) > x > y > \mathbf{LS}(H)$ for some numbers x, y, then $G - H > x - y$ and $x - y$ is bigger than any infinitesimal. Thus, we may assume that $\mathbf{RS}(G) = \mathbf{LS}(H) = x$. By the Number-Avoidance Theorem (Theorem 6.12), we have $\mathbf{RS}(G - x) = 0$ and $\mathbf{LS}(x - H) = 0$; therefore, $G - x \geq \epsilon$ and $\delta \geq H - x$ for some infinitesimals ϵ, δ. Combining these results, $G - H \geq \epsilon - \delta$.   □

This leads to a result that we will use often; see [NO11].

Corollary 10.6. *Let a and b be numbers with $a \geqslant b$. Then, $a \geqslant_\epsilon \{a \mid b\} \geqslant_\epsilon b$.*

For example, 3 and $\{3 \mid 0\}$ are incomparable, but $3 \geq_\epsilon \{3 \mid 0\}$ since $3 - \{3 \mid 0\} + {\uparrow} > 0$.

With slight tweaks to the *-ish* definition, there are the same reductions as for canonical form.

Definition 10.7. Let G be any game.

- A left option G^L is *Inf-dominated* if $G^L \leq_\epsilon G^{L'}$ for some other left option $G^{L'}$.

- A left option G^L is *Inf-reversible* through G^{LR} if $G^{LR} \leq_\epsilon G$. Just as with reversibility on page 91, we call G^{LRL} the *replacement set*.

The definitions for right options are similar.

Dominated options in *-ish* require one caveat: that G is not a number.

Lemma 10.8. *If G is not a number and G' is obtained from G by eliminating an Inf-dominated option, then $G' \sim_\epsilon G$.*

Theorem 10.9. *If $G = \left\{\mathcal{G}^L \mid \mathcal{G}^R\right\}$ is not a number and $G' = \left\{\mathcal{G}^{L'} \mid \mathcal{G}^{R'}\right\}$ is a game with $\mathcal{G}^{L'} \sim_\epsilon \mathcal{G}^L$ and $\mathcal{G}^{R'} \sim_\epsilon \mathcal{G}^R$, then $G' \sim_\epsilon G$.*

Naturally, in the above theorem, the relation "$\sim_\epsilon$" of sets means that sets are the same size and can be paired up with "$\sim_\epsilon$" games.

Example 10.10.

- Let $F = \{\{1 \mid 1\} \mid \{-1 \mid -\{-1 \mid -2\}\}\} = \{1* \mid -1\!+_1\}$. Since F is not a number, its left option can be replaced by 1 and its right option by -1; thus, $F \sim_\epsilon \{1 \mid -1\}$. However, if $F = \{0 \mid 1+ \,_1\}$ then $F = 1$, but replacing $1+ \,_1$ by 1 would give the game $\{0 \mid 1\} = \frac{1}{2}$. This explains why we needed the caveat "if G is not a number" in Theorem 10.9.

- Let $G = \{1, \{1|0\}|0\}$. Since $1 \geq_\epsilon \{1|0\}$ then $\{1|0\}$ is an Inf-dominated left option of G, which yields $G \sim_\epsilon \{1|0\}$.

- Let $H = \{2{\uparrow}|2{\downarrow}\}$ and note that $H \neq 2$ since the left option is greater than the right option. Now, $2{\uparrow} \sim_\epsilon 2{\downarrow} = 2$, which gives $H \sim_\epsilon \{2 \mid 2\} = 2*$. Since $2* \sim_\epsilon 2$, finally we see that $H \sim_\epsilon 2$.

- Let $K = \{2*|2{\downarrow}\}$, and again $K \neq 2$ but still $K \sim_\epsilon 2$. Now, $H \sim_\epsilon K$ but $H - K$ is a first-player win.

Bypassing reversible options requires more care.

Lemma 10.11. *If G is not number-ish and G_0 is obtained from G by bypassing an Inf-reversible option, then $G_0 \sim_\epsilon G$.*

For example, let $G = \{2, \{3 \mid *\} \mid 0\}$. The left option $\{3 \mid *\}$ is Inf-reversible since $G - * \geq_\epsilon 0$. (Left wins $G - * + {\uparrow}$ going second.) The replacement set is 0 and $G \sim_\epsilon \{2, 0 \mid 0\} = \{2 \mid 0\}$.

If G is number-ish then Inf-reversible moves cannot be bypassed. For example, let $G = \,_2 = \{0 \mid \{0 \mid -2\}\}$. Since $\mathbf{LS}(G) = \mathbf{RS}(G) = 0$, $G \sim_\epsilon 0$. Now $\{0 \mid -2\}$ is reversible through G^{RL} since $0 < \,_2$. The replacement set is $0^R = \emptyset$, and bypassing $\{0 \mid -2\}$ results in the game $\{0 \mid\} = 1 \neq_\epsilon G$.

Definition 10.12. A game G is said to be in *reduced canonical form* provided that, for every follower H of G, either

- H is a number in canonical form or

- H is not a number or a number plus an infinitesimal and has no Inf-dominated or Inf-reversible options.

The reduction theorems show that every game is $\sim_\epsilon$ to one that is in reduced canonical form. That is:

Theorem 10.13. *For any game G, there is a game $\mathrm{rcf}(G)$ in reduced canonical form such that $G \sim_\epsilon \mathrm{rcf}(G)$.*

With some extra work [GS07, Sie13], it can also be shown that this form is unique. That is:

Theorem 10.14. *Suppose that G and H are in reduced canonical form. If $G \sim_\epsilon H$, then G and H are identical.*

Example 10.15. What is the reduced canonical form of

$$G = \{\{2 \mid 1\}, \{3 \mid 0\}, 2 \mid *\}?$$

First, $G > 0$. Second, if G is a number x then, in particular, $2 \triangleleft x \quad *$, which is impossible. By Corollary 10.6, $2 \geq_\epsilon \{2 \mid 1\}$ and certainly $* \sim_\epsilon 0$; therefore, by Lemma 10.8 and Theorem 10.9, $G \sim_\epsilon \{\{3 \mid 0\}, 2 \mid 0\}$. Finally, since G is not number-ish ($LS(G) = 2$ and $RS(G) = 0$) and $G > 0$, then $\{3 \mid 0\}$ reverses out to give $G \sim_\epsilon \{2 \mid 0\}$.

10.3 Hereditary-Transitive Games

CRICKET PITCH and MAZE both have the property that if a player were allowed to make two moves in succession, the resulting position could have been reached in one move. We shall see that games with this characteristic have particularly simple reduced canonical forms. First though, we should define them.

Definition 10.16. A game G is *hereditary-transitive* (HT) if $H^{\mathcal{LL}} \subseteq H^{\mathcal{L}}$ and $H^{\mathcal{RR}} \subseteq H^{\mathcal{R}}$ for all followers H of G.

In the literature, hereditary-transitive games are sometimes called *option-closed*.

Exercise 10.17. Explain why any position, G, of MAZE is hereditary-transitive.

Theorem 10.18. *Let G be a hereditary-transitive game. The reduced canonical form of G is either a number or a switch, $\{a \mid b\}$, where a and b are numbers.*

Proof: As (almost) always we proceed by induction. Let G be hereditary-transitive, and suppose that the result holds for all simpler hereditary-transitive games (in particular, for all of G's options). If G is a number, there is nothing to prove. If G is not a number, the reduced canonical forms of all the left options of G are all switches or numbers. By Theorem 10.9, since G is not a number, any left option can be replaced by its reduced canonical form. If the switch $\{a \mid b\}$ is the reduced canonical form of a left option, then so is a (since G is hereditary-transitive). Now, by Corollary 10.6, $a \geq_\epsilon \{a \mid b\}$, and by Lemma 10.8 and Theorem 10.9, the switch is Inf-dominated and can be removed. Thus, all the left options of the reduced canonical form of G are numbers, and, of course, there can only be one such (the rest will be dominated). As the same applies to the right options, the final form is either a switch or a number.   □

The importance of this result is that if you did the exercises in Section 5.5, then you know how to determine the winner (or outcome class) of a game that is the sum of numbers and switches. So, as a first step in analysing a sum of hereditary-transitive games, you could compute their reduced canonical forms. If their sum is positive or negative, then you know the outcome type of the sum of the original games. Only if the sum is of type $\mathcal{P}$ or $\mathcal{N}$ might you be unsure, since then the exact infinitesimals that the reduced canonical form has suppressed could play a role.

10.4   Maze

The algorithm for finding reduced canonical forms of positions in MAZE is very simple. Basically, it amounts to filling in the cells of the entire MAZE with their reduced canonical forms, starting from the terminal cells. The main point is that if $\{k \mid j\}$, $k \geq j$, is an option of cell C for Left, then k is also an option. Since $k \geq_\epsilon \{k \mid j\}$, the switches can be ignored when calculating the reduced canonical form of C.

Algorithm 10.19. (Labelling Maze Cells)

- Start: No cell is labelled.

- Integers: Label all the cells that have no left or right options with appropriate integer values.

- Update: When there is an unlabelled cell, chose a cell, C, all of whose options have been labelled:

  1. $\ell = \max\{k : k$ is a number and is the label of a left option of $C\}$;

  2. $r = \min\{j : j$ is a number and is the label of a right option of $C\}$;

  3. Label C with $\{\ell \mid r\}$, unless $\ell = r$ then label C with ℓ.

- Output: The label on a cell is the reduced canonical form of the cell.

Example 10.20. Consider this MAZE position:

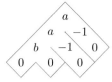

The values of 0 (nobody has a move) and -1 (Right has exactly one move) arise in the "Integers" phase and are easy to understand. The reduced canonical form of the value in the cell labelled b is 0 (actual value is $*$) since both Left and Right have just one move, that to 0. For the cells labelled a, Left has a move to the number 0 (the move to b or a is ignored), and Right has a move to both 0 and -1 and the latter dominates. Therefore, the reduced canonical form of the value in these cells is $\{0 \mid -1\}$.

10.5 Ordinal Sum

The ordinal sum construction was first noticed in HACKENBUSH when played on a tree. Play can occur in the leaves, but once a move is made on the trunk, no further leaf play is possible. The important part of the game is the trunk, but the leaves make a contribution. The game represents a type of *ordinal sum:* given games $G_1, G_2, \ldots, G_n$, a player may choose to move in, say, G_i but, after this move, neither player can ever play in G_j, $j > i$.

Conway [Con01] developed a theory of ordinal sums where the base game was restricted to be in canonical form. Neil McKay [McK16] generalized this theory to allow for a general base, not necessarily in canonical form. The analyses here admit non-canonical base games, so we present McKay's definition.

Definition 10.21. The *ordinal sum* of games G and H, written $G : H$, is the game

$$G : H \stackrel{\text{def}}{=} \left\{ G^{\mathcal{L}}, G : H^{\mathcal{L}} \mid G^{\mathcal{R}}, G : H^{\mathcal{R}} \right\}.$$

We refer to G as the *base* and H as the *branch*.[1]

Conway's definition is the same except for an additional requirement that the base, G, be in canonical form. With either definition, the ordinal sum does not depend on the second game being in canonical form since it is easy to see that, for instance, a dominated left option H^L of H yields a dominated option $G : H^L$ of $G : H$.

Since neither player can play in 0, $G : 0 = G$ and $0 : G = G$. Nimbers combine nicely: $* : * = \{0, * : 0 \mid 0, * : 0\} = \{0, * \mid 0, *\} = *2$, and in general $*n : *m = *(n+m)$.

---

[1]McKay called the branch the *subordinate,* but we prefer the horticultural theme.

Exercise 10.22. Confirm that $*n : *m = *(n+m)$.

Fractions are a little less friendly: for instance,

$$\frac{1}{2^p} : \frac{1}{2^q} = \frac{1}{2^p} + \frac{1}{2^{p+q+1}}.$$

(See Problem 1.)

There is a difference between the two definitions. With McKay's definition

$$\{*2 \mid *2\} : * = \{*2, \{*2 \mid *2\}:0 \mid *2, \{*2 \mid *2\}:0\}$$
$$= \{*2, \{*2 \mid *2\} \mid *2, \{*2 \mid *2\}\}$$
$$= \{*2, 0 \mid *2, 0\} = *;$$
$$\{*2 \mid *2\} : *2 = \{*2, \{*2 \mid *2\}:0, \{*2 \mid *2\}:* \mid *2, \{*2 \mid *2\}:0, \{*2 \mid *2\}:*\}$$
$$= \{*2, 0, * \mid *2, 0, *\}$$
$$= *3.$$

With Conway's definition, which requires the base to be in canonical form, $\{*2 \mid *2\} : * = 0 : * = *$ (no difference), but $\{*2 \mid *2\} : *2 = 0 : *2 = *2$.

Most often in a game, the positions are not in canonical form because there are dominated and reversible options. McKay's definition is more useful because it is applicable to these games. Siegel [Sie13] also follows this approach. We will present STIRLING-SHAVE in Section 10.6 as an example where we can make good use of ordinal sums to compute canonical forms.

Example 10.23. Table 10.1 gives examples of $a:b$ where a and b are integers.

$a\backslash b$	-2	-1	0	1	2
-2	-4	-3	-2	$-\frac{3}{2}$	$-\frac{5}{4}$
-1	-3	-2	-1	$-\frac{1}{2}$	$-\frac{1}{4}$
0	-2	-1	0	1	2
1	$\frac{1}{4}$	$\frac{1}{2}$	1	2	3
2	$\frac{5}{4}$	$\frac{3}{2}$	2	3	4

Table 10.1. Ordinal sums, $a:b$, of integers.

Example 10.24. Let $H = \{1 \mid 0\}$. If $K = -1$ then

$$H : K = \{1 \mid 0\} : \{\cdot \mid 0\}$$
$$= \{1 \mid 0, \{1 \mid 0\}:0\}$$
$$= \{1 \mid 0, \{1 \mid 0\}\}$$
$$= \{1, \{1 \mid 0\} \mid 0, \{1 \mid 0, \{1 \mid 0\}\}\}.$$

If $K = *$ then

$$H:K = \{1 \mid 0\}:\{0 \mid 0\}$$
$$= \{1, \{1 \mid 0\}:0 \mid 0, \{1 \mid 0\}:0\}$$
$$= \{1, \{1 \mid 0\} \mid 0, \{1 \mid 0\}\}.$$

It is not too surprising that putting two games on the same base game does not change the order relationship between them.

Observation 10.25. (The Colon Principle) Let G, H, and K be games. If $H \geq K$ then $G:H \geq G:K$. In particular, if $H = K$ then $G:H = G:K$ [BCG01, p. 219].

A second important feature of ordinal sums is that if $R = S:T$ then S can be used as an approximation for R when comparing against most, *but not all,* other games.

- If G and H are positions, $-(G:H) = (-G):(-H)$.

- $o(G:H) = o(G)$ if $o(G) \neq \mathcal{P}$, otherwise $o(G:H) = o(H)$.

- *Norton's Lemma* [Con01, p. 91]: Let G, H, and X be games with G not a follower of X. Then, $o(G - X) = o((G:H) - X)$.

- Ordinal sum is associative: $(G:H):J = G:(H:J)$.

The special case $1:G = \{0, 1:G^L \mid 1:G^R\}$ has particular importance in the analysis of TOPPLING DOMINOES.

Theorem 10.26. *Let $x > 0$ be a number. Then, $x = 1:y$ for some simpler number y.*

The theorem was originally proved in the context of BLUE-RED HACKEN-BUSH strings, where it was used to show that every such string encodes a unique number [BCG01].

Proof: It is easily seen that, for any integer n, $n + 1 = 1:n$ and $2^{-n} = 1:-n$, both of which satisfy the simplicity conclusion. So, assume that $x = \{x^L \mid x^R\}$ is in simplest form, with $x^R > x^L > 0$. By induction, $x^L = 1:y^L$ and $x^R = 1:y^R$ for some y^L and y^R that are respectively simpler than x^L and x^R. Let $y = \{y^L \mid y^R\}$. Then,

$$1:y = \{0, 1:y^L \mid 1:y^R\} = \{0, x^L \mid x^R\} = x,$$

since 0 is dominated by x^L. Since y^L and y^R are simpler than x^L and x^R, respectively, y is simpler than x. $\square$

Corollary 10.27. *The set of numbers is generated from $\{0\}$ by the transformations $x \mapsto -x$ and $x \mapsto 1:x$.*

10.6   Stirling-Shave

A position in the game STIRLING-SHAVE is a sequence of distinct positive integers, and a legal move is to remove any number in the sequence that is less than all of the numbers to its right, along with all of the numbers to its right. For instance, the legal moves from $[3,4,2,7,9,8]$ are to $[3,4]$, $[3,4,2]$, and $[3,4,2,7,9]$. Since STIRLING-SHAVE is an impartial game, we only need to compute nim-values to analyze it.

Determining the nim-value of a single STIRLING-SHAVE row is made easier by realizing that the ordinal sum is involved. For example, the game $G = [3,4,2,7,9,8]$ considered above can be regarded as $H : K = [3,4,2] : [7,9,8]$ because (i) the (only) move in H leaves $[3,4]$ and K has been removed and (ii) the moves in K to $[7,9]$ or $[]$ correspond to moves leaving $H : [7,9]$ or H. This observation is used for the reduction to ordinal sums.

Lemma 10.28. *Let* $G = [a_1, a_2, \ldots, a_n]$ *be a* STIRLING-SHAVE *position, and let* $a_k = \min\{a_i : i = 1, 2, \ldots, n\}$. *Then,* $G = [a_1, a_2, \ldots, a_k] : [a_{k+1}, a_{k+2} \ldots, a_n]$.

Proof: Let $H = [a_1, a_2, \ldots, a_k]$ and $K = [a_{k+1}, a_{k+2} \ldots, a_n]$. In G, the leftmost pile that can be removed is a_k. In $H : K$, this corresponds to making the only legal move in H and eliminating K.

If a_i, $i > k$, is a legal move in G, then there is a legal move in K to $K' = [a_{k+1}, a_{k+2} \ldots, a_{i-1}]$ and the move from $H : K$ to $H : K'$ is also legal (and all legal moves in the branch of $H : K$ are of this type).

So, the legal moves in G and in $H : K$ directly correspond with one another, and hence by induction $G = H : K$.   □

Clearly, Lemma 10.28 can be used recursively to evaluate any game of STIRLING-SHAVE, but we can also consider breaking G down further in a single pass through the position — insert a colon after the minimum, then after the minimum of the remaining sequence, and so on, until we reach the right-hand end.

Definition 10.29. *In the* STIRLING-SHAVE *position* $[a_1, a_2, \ldots, a_n]$, *the hole numbers are* $a_{j_1}, \ldots, a_{j_\ell}$ *defined recursively, where* $a_{j_1} = \min\{a_1, a_2, \ldots, a_n\}$ *and, for* $i > 1$, $a_{j_i} = \min\{a_{j_{i-1}+1}, \ldots, a_n\}$.

The hole numbers are also called *right-to-left minima* since they correspond to the minimum values seen when we read the sequence from right to left. In particular, the last hole number of $[a_1, a_2, \ldots, a_n]$ will always be a_n.

Corollary 10.30. *Let* $G = [a_1, a_2, \ldots, a_n]$ *be a* STIRLING-SHAVE *position, and let the hole numbers be* $a_{j_1}, \ldots, a_{j_\ell}$. *Then,*

$$G = [a_1, a_2, \ldots, a_{j_1}] : ([a_{j_1+1}, \ldots, a_{j_2}] : (\cdots : [a_{j_{\ell-1}}, \ldots, a_n]) \ldots).$$

In this decomposition the ordinal sums are evaluated from the right first. This allows us to drop the parentheses *if we agree to always evaluate in the right-to-left order*. Evaluating in a different order may give a different result. Notice that each of the sequences appearing in the decomposition ends with its minimum element (i.e., there is only one hole number) and corresponds to a game in which each player has only one move. For a sequence A let $\{A\}$ denote the impartial game having a single left and right option, which is the STIRLING-SHAVE position A.

In $G = [3,4,2,7,9,8]$, the hole numbers are 2, 7, and 8 and, as an ordinal sum, $G = [3,4,2]:[7]:[9,8]$. However, using the observation above, it is also the case that

$$G = \{[3,4]\}:\{[]\}:\{[9,8]\} = \{[3,4]\}:*:\{[9]\}.$$

More generally:

Corollary 10.31. *Let $G = [a_1, a_2, \ldots, a_n]$ be a STIRLING-SHAVE position, and let the hole numbers be $a_{j_1}, \ldots, a_{j_\ell}$. Then,*

$$G = \{[a_1, a_2, \ldots, a_{j_1-1}]\}:\{[a_{j_1+1}, \ldots, a_{j_2-1}]\}:\cdots:\{[a_{j_\ell+1}, \ldots, a_{n-1}]\}.$$

Example 10.32. The hole numbers of $G = [4,1,3,7,2,6,5]$ are 1, 2, and 5 and, doing one hole number at at time,

$$G = \{[4]\}:\{[3,7]\}:\{[6]\}.$$

If we want to find the value of a STIRLING-SHAVE position and we know the values of "simpler" positions, then we will know the nim-value $*k$ of the final term in its representation as an ordinal sum and also the nim-value $*n$ of the second-to-last term (when we use the representation in Corollary 10.30). To combine these values, we need to evaluate $\{*n \mid *n\}:*k$. This is accomplished by extending the mex rule for impartial games.

Definition 10.33. The mex *rule of order k* of a set S of non-negative integers, written $\mathrm{mex}_k(S)$, is defined recursively:

$$\mathrm{mex}_0(S) = \mathrm{mex}(S);$$

$$\mathrm{mex}_k(S) = \mathrm{mex}(S \cup \{\mathrm{mex}_0(S), \ldots, \mathrm{mex}_{k-1}(S)\}).$$

For example, if $S = \{0,3,4,6\}$, then

$$\mathrm{mex}_k(S) = \begin{cases} 1 & \text{if } k = 0, \\ 2 & \text{if } k = 1, \\ 5 & \text{if } k = 2, \\ 7 + (k-3) & \text{if } k \geqslant 3. \end{cases}$$

Practically, $\mathrm{mex}_k(S)$ is equal to $\mathrm{mex}(S')$ where

$$S' = S \cup \{k \text{ least non-negative integers not included in } S\}.$$

Working with mex_k is essentially the same as working with mex.

Lemma 10.34. *Let S be a set of non-negative integers, let k be a non-negative integer, and let $h = \mathrm{mex}_k(S)$. For any integer j, $0 \le j < h$, either $j \in S$ or there exists $\ell < k$ such that $j = \mathrm{mex}_\ell(S)$.*

Proof: Since h is the least non-negative integer not in

$$\mathrm{mex}(S \cup \{\mathrm{mex}_0(S), \dots, \mathrm{mex}_{k-1}(S)\})$$

and since $j < h$, we have that j is one of the elements of

$$\mathrm{mex}(S \cup \{\mathrm{mex}_0(S), \dots, \mathrm{mex}_{k-1}(S)\}). \qquad \square$$

Theorem 10.35. *Let G and K be impartial games (not necessarily in canonical form), and let $k = \mathcal{G}(K)$. Then,*

$$\mathcal{G}(G\!:\!K) = \mathrm{mex}_k(\mathcal{G}(G') \,:\, G' \text{ is an option of } G).$$

Proof: The proof will proceed by induction on k. The base case, $k = 0$, follows from Theorem 7.7.

Let $S = \{\mathcal{G}(G') \mid G' \text{ is an option of } G\}$. As usual, to prove games equal, we will prove that $* \mathrm{mex}_k(S) - (G\!:\!*k) = 0$, but since all the games are impartial, this is the same as proving $* \mathrm{mex}_k(S) + (G\!:\!*k) = 0$.

Suppose that the first player chooses an option of $* \mathrm{mex}_k(S)$; call it $*j$. By Lemma 10.34 there are two possibilities. The first is $j \in S$, which can only happen if $*j = *\mathcal{G}(G')$ for some option G' of G and the second player responds by moving from $G\!:\!K$ to G' and $*j + *\mathcal{G}(G') = 0$. The second possibility is $j = \mathrm{mex}_\ell(S)$ for some $\ell < k$. Since $\mathcal{G}(K) = k$ there is an option K' with $\mathcal{G}(K') = \ell$. By induction, $* \mathrm{mex}_\ell(S) = G\!:\!K'$; therefore, the second player wins by moving to $* \mathrm{mex}_\ell(S) + G\!:\!K'$.

Suppose that the first player moves in $G\!:\!K$. If the first player moves in K to K' where $\mathcal{G}(K') > \mathcal{G}(K)$, then the second player plays to K'' where $\mathcal{G}(K'') = \mathcal{G}(K)$ and the result follows by induction. (This shows why we can assume that K is in canonical form.) If the first player moves in K to K' where $j = \mathcal{G}(K') < \mathcal{G}(K)$ or to G', an option of G, then the second player moves in the other component to, respectively, either $* \mathrm{mex}_j(S)$ or $*\mathcal{G}(G')$. In both cases, by induction, the resulting game is 0. $\qquad \square$

Let's make Theorem 10.35 explicit for the particular types of games that result in the ordinal decomposition of STIRLING-SHAVE:

Corollary 10.36. *Let* $m, k \geqslant 0$. *Then,*

$$\{*m \mid *m\} : *k = \begin{cases} *k & \text{if } k < m, \\ *(k+1) & \text{if } k \geqslant m. \end{cases}$$

Example 10.37. In the game G, the hole numbers are indicated by bars:

$$G = [\overline{1}, \overline{2}, \overline{3}, \overline{4}, 16, 17, 18, 19, 20, 21, 22, \overline{5}, 12, 13, 14, 15, \overline{6}, \overline{7}, 23, 24, \overline{8}, \overline{9}, 25, \overline{10}, 26, \overline{11}].$$

In order to simplify our calculations, we first identify the game value of each block:

$$\underbrace{\overline{1}}_{*}, \quad \underbrace{\overline{2}}_{*}, \quad \underbrace{\overline{3}}_{*}, \quad \underbrace{\overline{4}}_{*}, \quad \underbrace{16,17,18,19,20,21,22,\overline{5}}_{\{*7 \mid *7\}}, \underbrace{12,13,14,15,\overline{6}}_{\{*4 \mid *4\}}, \underbrace{\overline{7}}_{*}, \underbrace{23,24,\overline{8}}_{\{*2 \mid *2\}}, \underbrace{\overline{9}}_{*}, \underbrace{25,\overline{10}}_{\{* \mid *\}}, \underbrace{26,\overline{11}}_{0}$$

Note that the sub-position $[26, \overline{11}]$ has value 0 since it is a second-player win.

Applying Corollary 10.36 sequentially, the nim-value of the above STIRLING-SHAVE row is 9:

Example 10.38. Let $G = [10, 2, 12, 11, 3, \overline{1}, 7, 9, 5, 6, \overline{4}, 8]$, where the hole numbers are indicated by bars. To evaluate G, first the values of both $[10, 2, 12, 11]$ and $[7, 9, 5, 6]$ have to be found.

Now,

$$[10, 2, 12, 11, 3] = [10, 2] : [12, 11, 3] = 0 : * = *$$

and

$$[7, 9, 5, 6] = [7, 9, 5] : [6] = \{*2\} : * = *2.$$

Decomposing G and evaluating gives

$$\begin{aligned} G &= \{[10, 2, 11, 11, 3, \overline{1}\} : [7, 9, 5, 6, \overline{4}] : [8] \\ &= \{[10, 2, 12, 11, 3]\} : \{[7, 9, 5, 6]\} : * \\ &= \{*\} : \{*2\} : * = \{*\} : * = *2. \end{aligned}$$

10.7   Even More Toppling Dominoes

A fact, discovered after the first edition went to print, is that every number appears as a TOPPLING DOMINOES position[2] and moreover that position is unique. The proof about numbers, which we follow here, and much more about the game can be found in [FNSW15]. That paper covers the more general game where green dominoes may also be present.

The outline of our argument is as follows:

1. We associate to each TOPPLING DOMINOES position G a new position $f(G)$ of value $1:G$. Since there is an obvious transformation $G \mapsto -G$ (replace blue dominoes with red ones and vice versa), we immediately obtain all numbers.

2. We then show that *every* positive number G is necessarily of the form $f(H)$ for some H.

3. We use a simple induction argument to establish uniqueness.

First, we explore the transformation f.

Definition 10.39. Let f be a function mapping strings to strings such that, for a TOPPLING DOMINOES string α, $f(\alpha)$ is obtained by inserting one additional L in every maximal substring of Ls in α, including empty substrings. For example, $f(L) = LL$ and $f(RRLRLL) = LRLRLLRLLL$.

Equivalently, f can be described by the following transformation, where $\wedge$ denotes the "start of string" marker for α:

$$\wedge \to \wedge L; \quad L \to L; \quad R \to RL.$$

Theorem 10.40. [FNSW15] *For any* TOPPLING DOMINOES *position G, we have* $f(G) = 1:G$.

Proof: Let $H = \{0, f(G^L) \mid f(G^R)\}$. It suffices to show that $f(G) = H$, for then the theorem follows by induction.

Step 1: We first show that every left (respectively right) option of H is a left (respectively right) option of $f(G)$. Consider a typical G^R. We may write $G = \alpha R \beta$ with $G^R = \alpha$ or β. By definition of f, we have $f(G) = f(\alpha)Rf(\beta)$, so $f(\alpha)$ and $f(\beta)$ are necessarily right options of $f(G)$.

Likewise, for any G^L we may write $G = \alpha L \beta$ with $G^L = \alpha$ or β. If $G = \alpha L \beta$, then $f(G) = f(\alpha)f(\beta)$, since $f(\beta)$ introduces a new L at the beginning. Since any string in the image of f both starts and finishes with an L, $f(\alpha)$ and $f(\beta)$ are left options of $f(G)$.

---

[2]Recall that a blue domino is represented by L and a red by R.

Finally, $f(G)$ has a move to 0, since it necessarily has a blue domino on the west end.

Step 2: To complete the proof, we show that for every option of T of $f(G)$, *either* T is an option of H *or* T reverses through 0.

First of all, inspection of Definition 10.39 reveals that *every* right option of $f(G)$ is obtained as in Step 1, since every red domino in $f(G)$ arises as the image of a red domino in G.

However, there are additional left options $T = f(G)^L$ that are not of the form 0 or $f(G^L)$. In particular, they are those that expose a red domino, but these moves reverse out through Right's moves to 0, since $0 < f(G)$.   □

Corollary 10.41. [FNSW15] *For every number x, there is a* TOPPLING DOMI-NOES *position $G = x$.*

Proof: Theorem 10.40 yields the mapping $G \mapsto 1 : G$ and reversing color yields the mapping $G \mapsto -G$, so this result follows from Corollary 10.27.   □

A few more lemmas are needed before we can prove uniqueness.

Lemma 10.42. [FNSW15] *Let G be a* TOPPLING DOMINOES *position with $G > 0$. If G does not contain RR as a substring, then there exists a* TOPPLING DOMINOES *position H such that $G = f(H)$.*

Proof: H is obtained by removing a single L from every maximal substring of Ls in G.   □

The next result is a corollary of a more general result found in [Con01]. In that result, "1" is replaced by a general game G, but then the three games have to satisfy a technical condition.

Lemma 10.43. [FNSW15] $1 : H \geq 1 : K$ *if and only if $H \geq K$.*

Proof: When playing $1 : H - 1 : K$, a move on the base 1 taking a component to 0 can never be winning unless the other component is already 0. Hence, winning play on $1 : H - 1 : K$ is tantamount to play on $H - K$ until the last two moves.   □

Lemma 10.44. [FNSW15] *If a* TOPPLING DOMINOES *position $G > 0$ has consecutive red dominoes, then G is not a number.*

Proof: First note that such a G must begin and end with L (otherwise it is not the case that $G > 0$). Suppose (for contradiction) that $G = L\alpha A B \beta L$ is a number where A and B are red dominoes. Consider Right's move to $G^R = L\alpha A$. Since $G^R \rhd G$, Left must have a winning move on $G^R - G$. Since $-G$ is a number and G^R is not, Left must have a winning move on the G^R

component. By Lemma 5.63 (with R and L interchanged), this move must be to topple west on $L\alpha A$, since any topple to the east would leave a position incomparable to or less than G and hence not be winning. But, such a move leaves a position of the form $\gamma A - G$, and then Right has a response to $-G < 0$, a contradiction. □

Lemma 10.45. [FNSW15] *Let G be a* TOPPLING DOMINOES *position. If G is a number and $G > 0$, then $G \cong f(H)$ for some H.*

Refer to Chapter 4 for a refresher on "$\cong$."

Proof: It follows from the lemma that G is in the image of f. □

Theorem 10.46. [FNSW15] *Let G and K be* TOPPLING DOMINOES *positions, and suppose that $G = K = x$, a number. Then, $G \cong K$.*

Proof: For $x = 0$ this is just Lemma 5.61 on page 130. We proceed by induction on x.

First, suppose that $x > 0$. By Lemma 10.44 neither G nor K can contain consecutive red dominoes and thus $G = f(H)$ and $K = f(H')$ for some H, H'. By Theorem 10.40, $G = 1:H$ and $K = 1:H'$. Now we also know that $x = 1:y$ for some number y simpler than x. By Lemma 10.43, we have that $H = H' = y$, so by induction $H \cong H'$. Therefore, $G \cong G'$.

The case $x < 0$ is symmetric. □

Corollary 10.47. *All* TOPPLING DOMINOES *positions that are numbers are palindromes.*

Proof: If K is the reversal of G, then certainly $K = G$, so by Theorem 10.46 we have $K \cong G$. □

Problems

1. Show the following:

    (a) If n is a positive integer then $1 : n = n + 1$, and if n is negative then $1 : n = \frac{1}{2^{|n|}}$.

    (b) If n is a positive integer then $1 : \frac{1}{2^n} = 1 + \frac{1}{2^n}$.

    (c) If p and q are positive integers then $\frac{1}{2^p} : \frac{1}{2^q} = \frac{1}{2^p} + \frac{1}{2^{p+q+1}}$.

2. Let n and m be non-negative integers. Show that $*n : *m = *(n + m)$.

3. Find the reduced canonical form of the MAZE positions given in Problem 13 of Chapter 5 on page 135.

4. STONES is played on a rectangular board with blue and red stones. Label the board $[a, b]$, $a, b \geq 0$. All moves are toward $[0, 0]$, the lower-left corner. The edge row and column are special. For either color stone, on the bottom edge, $[0, b]$, Right can move it like a rook, and on the left edge, $[a, 0]$, Left can move it like a rook. Elsewhere, both players can move a stone diagonally. For a blue stone, Left can remove it from the board at any time, and Right can remove it only if $a < b$. The moves are reversed for a red stone. More than one stone can occupy a square.

(a) Find the value of a blue stone on $[a, b]$, $a \geq b$.

(b) Find the reduced canonical form of a blue stone on $[a, b]$, $a < b$.

(c) Who wins  ?

Preparation for Chapter ω

Where to from here? We hope you appreciate that the study of combinatorial games is by no means a closed book. In this chapter we will describe some areas where research is ongoing. There are many others — we hope that you are tempted to explore either some of the ones we'll describe here or other topics that interest you.

To the instructor: As noted this is not a chapter of "content" but simply provides a taster of where students might like to explore further. Some of these could well be used for research project topics.

Chapter ω
Further Directions

An optimist, in the atomic age, is a person
who thinks the future is uncertain.

Howard Lindsay and Russell Crouse in
State of the Union

After having read this text, we hope you are eager to learn more. The first
resource you should use remains the gem *Winning Ways* [BCG01], which (as
you might have guessed from the number of times we referenced it) remains
authoritative and thorough.

READ IT!

Having said that, combinatorial game theory is an active field of current research, and the most recent innovations are not covered in detail in
WW [BCG01], but many can be found in Aaron Siegel's comprehensive text
Combinatorial Game Theory [Sie13], which you should read after finishing our
book. In this chapter, we indicate some other directions for further reading,
with an emphasis on newer results. While making no attempt to be complete,
we hope to provide a flavor for what we could not cover in this text.

ω.1 Transfinite Games

Games with ordinal birthdays are quite natural. For example, play the following
game on polynomials with non-negative coefficients. (You are asked to analyze
this game in Problem 7 of Chapter 7.) A move is to choose a single polynomial
and reduce one coefficient and arbitrarily change or leave alone the coefficients
on the smaller powers — $3x^2 + 15x + 3$ can be reduced to $0x^2 + 19156x + 2345678 = 19156x + 2345678$. If there are no powers of x^i for $i > 0$, then this

is just NIM; if there are powers of x, then this is NIM with heaps of ordinal heights. Thus, $1x^1 = \{0, *, *2, *3, \ldots \mid 0, *, *2, *3, \ldots\}$, which is born on day ω.

With the correct notion of ordinal powers of 2, the theory for finite NIM extends to heaps of transfinite ordinals. Similarly, it is not at all hard to imagine playing HACKENBUSH on infinite graphs. For example, a string of blue edges might be

$$\cdots = \{0, 1, 2, \ldots \mid \} = \omega,$$

the first infinite ordinal; adding a single blue edge after the "$\cdots$" would give $\{0, 1, 2, \ldots, \omega \mid \} = \omega + 1$; changing the added edge to red gives $\{0, 1, 2, \ldots \mid \omega\}$, which behaves like $\omega - 1$, a value that does not occur in the standard set-theoretical development of ordinal arithmetic. Taking a single blue edge and adding an infinite string of red edges gives $\{0 \mid 1, \frac{1}{2}, \frac{1}{4}, \frac{1}{8}, \ldots\} = \frac{1}{\omega}$. Note that, in these examples, *the elimination of all dominated options is no longer a safe operation;* in the latter case each and every option $\frac{1}{2^n}$ is dominated by another option $\frac{1}{2^{n+1}}$, but the original game is not equivalent to $\{0 \mid \} = 1$.

The HACKENBUSH string

$$\cdots = \frac{1}{3},$$

simply because in binary $\frac{1}{3} = .010101\ldots$. Similarly, all real numbers can be expressed in binary (with up to ω bits to the right of the decimal point) and are born by day ω. With a little imagination, it is not too difficult to find a game that should have value $\frac{\omega}{2}$, but more exotic values such as $\omega^{\frac{1}{2}}$ and $\omega^{\frac{1}{\omega}}$ also exist. More can be found in *On Numbers and Games* [Con01] and [Con02], *Combinatorial Game Theory* [Sie13], or in the more light-hearted exposition *Surreal Numbers* [Knu74].

ω.2 Algorithms and Complexity

In PARTIZAN ENDNIM (Section 2.6), it seemed that one might have to analyze most of the game tree to actually find a good move. But, in that instance we were able to provide a relatively simple algorithm to determine the outcome class of a position, and hence a good move. On the other hand, in Theorem 5.54, we saw that the disjunctive sum of relatively simple-looking games could be hard to calculate. Determining the value of a CLOBBER position with just one red piece (on an arbitrary graph) [AGNW05] and also determining whether a player can jump and finish the game PHUTBALL in a single move [DDE02] are both NP-complete problems. In fact, there is a general, but completely imprecise, observation that any game that is actually interesting to play throws up algorithmically difficult problems at every turn.

In general, one would expect strategies and values to be complicated to find since we are looking for the resolution of a satisfiability problem for a formula that contains a long string of alternating quantifiers; for example, Left might want

$$\exists G^L \forall G^{LR} \exists G^{LRL} \forall G^{LRLR} \exists G^{LRLRL} \forall G^{LRLRLR} \ldots \exists G^{LR\ldots L} \geq 0.$$

In words, this simply says that there exists a left move, such that no matter what Right does there exists a left move, such that no matter what Right does there exists a left move, such that ..., such that Left wins. A (possibly) interesting question is whether or not there are simple conditions that determine classes of games whose values are easy to determine.

Combinatorial game theory starts at the end of the game and works back to the beginning, finding exact values when possible or at least outcome classes. Computer scientists are interested in algorithms and heuristics that allow play to start at the beginning and guide the player through to the end of the game. If we have a complete theory, then both should coincide. However, even then, if the evaluation takes too long to compute, then how does one proceed? This leads directly to the problem of finding strategies and heuristics for play, which may not be optimal but are rarely bad. We mentioned a few such in Chapter 1 and other based on temperature considerations in Chapter 8.

If you are interested in this area, then here are good websites from which to start:

- The Games Group at the Department of Computer Science of the University of Alberta, http://www.cs.ualberta.ca/~games/;

- Erik Demaine's games pages, http://erikdemaine.org/games/; and

- Aviezri Fraenkel's page http://www.wisdom.weizmann.ac.il/~fraenkel/ and, in particular, the papers [Fra00, Fra04].

ω.3   Loopy Games

The theory presented in this book assumes that no position may be repeated; otherwise, the game may not terminate, violating one of the restrictions that we have placed on the games. However, there are many loopy games, so called because their game graphs may contain cycles, for example, FOX & GEESE, HARE & HOUNDS, BACKSLIDING TOADS & FROGS, and CHECKERS. Their theory is more complicated and difficult. There is not, as yet, a good general definition of canonical form. However, in some loopy games, even though there is the potential for infinite play, one player has a winning strategy, where the game terminates in a finite number of moves, such that he never needs to avail himself of the repetitions. For these games, much of the *finite* theory can be mimicked. A game G is called a *stopper* if there is no infinite alternating sequence of play from any follower of G. That is, the game graph of G contains no alternating cycles. Stoppers, when played in isolation, are guaranteed to terminate, but infinite play might be possible if G is part of a disjunctive sum. In the 1970s, Conway, Bach, and Norton showed that every stopper has a unique canonical form. Aaron Siegel extended the theory, finding a general method to determine whether an arbitrary loopy game is equivalent to a stopper [Sie07, Sie05, Sie13]. However, loopy games in general are not well understood.

ω.4   Kos: Repeated Local Positions

GO is a peculiar game: the *ko* rule forbids most repeated positions so the game is not inherently loopy, but local positions may repeat arbitrarily often. Even the game of WOODPUSH (played on a finite strip of squares with blue and red pieces; a piece retreats to the next empty square, Left to the left and Right to the right, and eventually moves off the strip, except that it can push a contiguous string of stones including at least one piece belonging to the opponent one space in the opposite direction) has many of the same ko situations that occur in GO:

$$\boxed{\ \ \bullet\bullet\bullet\ } \overset{L}{\dashrightarrow} \boxed{\ \ \ \bullet\bullet\bullet} \overset{R}{\dashrightarrow} \boxed{\ \bullet\bullet\ \bullet} \overset{L}{\dashrightarrow} \text{ko-threat}$$

$$\overset{R}{\dashrightarrow} \text{answers ko-threat} \overset{L}{\dashrightarrow} \boxed{\ \ \ \bullet\bullet\bullet\ } \overset{R}{\dashrightarrow} \boxed{\ \ \bullet\bullet\bullet\ }$$

Note that, in the global game, the final position is not the same as the original since there have been two moves, the ko-threat and the answer, played elsewhere. The position $\boxed{\ \bullet\bullet\ }$ is hot, but how hot? The thermographs for such situations have to take into account that one opponent might get two consecutive moves, since a player may ignore a ko-threat somewhere else. The thermographs presented in this book only have slopes of ± 1 or ∞. The *extended thermographs* [Spi01, Sie13] required for GO and WOODPUSH can have slopes of $\pm\frac{1}{n}$ for any $n \neq 1$.

ω.5   Top-Down Thermography

Building a thermograph from the bottom up is time consuming. In practice, the shape of the thermograph near the temperature of the game is most important. Top-down thermography is the art (at the time of writing, it is not yet a science) of finding the slopes just before the thermograph turns into a mast.

For example, for the WOODPUSH position $\boxed{\ \bullet\bullet\ }$ we would have

$$-4 = \underset{(-4,0)}{\boxed{\bullet\ \ }} \leftrightarrow \underset{(-2,2)}{\boxed{\bullet\bullet\ }} \leftrightarrow \underset{(0,2)}{\boxed{\ \bullet\bullet\ }} \leftrightarrow \underset{(2,2)}{\boxed{\ \bullet\bullet\ }} \leftrightarrow \underset{(4,0)}{\boxed{\ \ \bullet}} = 4,$$

where the mean values and temperatures are written underneath. In this example, the temperature is the difference in numbers divided by the number of steps, i.e., $(4 - (-4))/4$, and the change in the mean value is obtained by moving in the adjacent positions and gaining the temperature. This is consistent across this example. The situation for ▢▢▢●●●▢ is much less clear.

ω.6   Enriched Environments

In cooling games, we have the concept of *taxing*. To use Elwyn Berlekamp's words, "No one likes paying taxes." He avoided taxation by introducing bonuses, or *coupons* [Ber96] (see also [Sie13]). In ENVIRONMENTAL GO, at the side of the board, there is a stack of coupons starting at 20 points and decreasing by $\frac{1}{2}$ until at the end there are three coupons each worth -1. On his turn, a player may move on the board or take a coupon — taking a coupon is also an answer to a ko-threat, and, at the end of the game, the value of the coupon is added to the points won on the board. The game is over when the last coupon is taken (assuming no resignation). Jiang Zhujiu and Rui Naiwei played the first ENVIRONMENTAL GO game in April 1998; see [Spi02].

The coupons reflect the temperature of the game on the board. If the temperature is low but the coupon value is high, both players will take coupons until, finally, a play on the board will gain more points than taking the next coupon.

This approach redefines temperature for numbers: the temperature essentially becomes the incentive. The temperature of 1 is -1, and the temperature of $\frac{3}{2}$ is $-\frac{1}{2}$.

ω.7   Idempotents

A game G is an idempotent if $G+G = G$. Idempotents have the effect of hiding from view games of smaller values. If there is an idempotent of the right size in play, then the players can restrict their attention to the positions with only the largest temperatures and values. For example, ✩ is an idempotent that essentially hides the nimbers and those infinitesimals that are infinitesimal with ↑, and it leads directly to the concept of atomic weights. Is there any equivalent notion for ⧾₂? Berlekamp [Ber02] lists and discusses other useful idempotents.

ω.8   Misère Play

In misère play, the last player to play loses. This seemingly innocuous change drastically affects the underlying theory. While [BCG01, Ch. 13] provides some rather helpful guidance, their chapter evidences how *hard* misère games are in general. As an exercise in the peculiarities of misère analysis, find two left-win games in misère play whose disjunctive sum is a right-win game in misère play. In misère play, if G has the property that $G+X$ has the same outcome as X for all games X, then G *can only be* the game $\{ \mid \}$. By contrast, in normal play, all second-player-win games are equivalent. In normal play, the equivalence classes were large, thereby providing a significant simplification in the analysis of games. If we try the same approach for misère games, it turns out that the equivalence classes are rather small.

Recently, however, Thane Plambeck had a brilliant insight that leads to a coherent theory for impartial misère games. In normal play, the values encountered are the stars 0, $*$, $*2$, etc., and these values give all the information needed to play in any disjunctive sum. The stars can also be regarded as elements of $\mathbb{Z}_2 \oplus \mathbb{Z}_2 \oplus \mathbb{Z}_2 \oplus \mathbb{Z}_2 \oplus \cdots$. Plambeck's insight is to construct semi-groups that replace $\mathbb{Z}_2$. One small limitation with this approach is that the information provided is only sufficient to play the game as a disjunctive sum with other positions of the same game.

Thane Plambeck and Aaron Siegel have made great progress, and we refer you to [SP08] or to the website http://www.miseregames.org/.

There is no "useful" theory that encompasses all partizan games since the equivalence classes are small and almost all games are incomparable. However, the same approach employed for impartial games has been used in partizan misère games. This has lead to work on subsets of partizan games based on structural properties:

1. *All-small games:* renamed *dicot* because "all-small" games are not infinitesimal under the misère convention; see [DRSS15].

2. *Placement games:* pieces are placed on a board but not moved or removed, for example, DOMINEERING.

3. *Dead-ending games:* if, in a position G, Left cannot move then Left cannot move in any follower of G, similarly for Right. This subset includes placement games. See [MR12].

Combinatorial Game Theory [Sie13] includes all of the impartial misère theory. Inspired by Plambeck and Siegel, partizan misère theory became a hot topic after publication of [Sie13].

ω.9 Scoring Games

Obviously, these are games that are won by the player with the highest score — the *score* is the difference between the Left score and the Right score — and who moves last has no bearing on the outcome of the game. Ties are now a possibility. The game $\langle 2 \mid -3 \rangle$ would be the game in which Left moves to a position with a score of 2 and Right to a position with score -3. We also need to consider positions like $\langle \emptyset^2 \mid \emptyset^{-3} \rangle$, a position with no moves but if it Left's turn then the score is 2 and if it is Right's then the score is -3. These positions occur when there are penalties or bonuses for running out, or running your opponent out, of moves. All of the games considered in this book would end with ^{0} entries. For example, the normal play games 0 and 1, let's call them $w(0)$ and $w(1)$, would be $w(0) = \langle \emptyset^0 \mid \emptyset^0 \rangle$ and $w(1) = \langle 0 \mid \emptyset^0 \rangle$; that is, Left has a move and Right doesn't but the final score is always 0.

The structure of scoring games is similar to that of misère games. Particularly, the equivalence classes are small and almost all games are incomparable. Even $w(0)$ and $w(1)$ are incomparable because in $\langle \emptyset^2 \mid \emptyset^{-3} \rangle + w(0)$ Left wins going first since the score will be 2, but she loses in $\langle \emptyset^2 \mid \emptyset^{-3} \rangle + w(1)$.

However, Ettinger [Ett96, Ett00] developed a theory for dicot scoring games, and Aaron Siegel [Sie15], in the context of misère games, developed a method to compare games to replace the normal play method of comparing the difference to 0. Building on these, Larsson, Nowakowski, Neto, and Santos [LaPNNS16] developed a theory for a large subclass of scoring games, called *guaranteed scoring games*. Placement games (mentioned in Section ω.8) are examples of games in this class

This is the largest class (in some sense) of games in which the order of normal play games is preserved. That is, in normal play $-1 < \downarrow < 0 < \uparrow < 1$, and this is still true when extended to guaranteed scoring games. However, it is not true when considering all scoring games.

Preparation for Appendix A

To the instructor: This appendix may be a bit of a stretch for the typical undergraduate. We highly recommend covering material up through Example A.5 on page 261. The remainder of the chapter provides additional motivation for a top-down view of induction and can safely be left for the stronger undergraduate to read at her leisure.

Appendix A
Top-Down Induction

Miss one step, and you will always be one
step behind.

Chinese proverb

A.1 Top-Down Induction

The type of induction most commonly used in combinatorial game theory (and
indeed in most of discrete mathematics) is somewhat different than that traditionally taught in beginning mathematics courses. One version of it is summarized below:

---

1. To prove a statement $P(s)$ for all $s \in S$, feel free at any time (i.e.,
   whenever convenient) to assume that $P(r)$ is true for all $r < s$.
   Whenever such an assumption is invoked write, "by induction."

2. Carefully review your proof, and prove any case not handled —
   "base case."

A few notes are in order:

- You may be concerned that this form of induction seems to be going
  "backward." Rather than saying, "Assume $P(n)$ and prove $P(n + 1)$"
  as you may have first learned, this says, "Prove $P(s)$ by using $P(r)$ for
  $r < s$." The two views are totally consistent; both use the smaller cases
  to help prove larger cases.

- You never have to decide in advance that you plan to use induction. You
  may use it whenever you wish in *any* proof, and later worry about any
  base cases you may have missed.

- *Most importantly,* S can be any set and $<$ can be any partial ordering that has the property that any non-empty subset of S has at least one minimal element.[1]

One of the advantages of this form of induction is that it often turns out that no base cases are necessary! Though it is true that the set S will contain minimal elements, these are often covered in the "by induction" step of the argument. This is because the argument that shows that "$P(s)$ is true provided that $P(r)$ is true for all $r < s$" may well not depend on there being any such r. Indeed, when such r do not exist, it may well be that $P(s)$ follows automatically without any further discussion — such statements are often referred to as being *vacuously true.*

The most elementary example of a partial ordering satisfying this minimality requirement is the non-negative integers under the usual $<$. In most proofs in this book, we usually have in mind the ordering of game positions, where $H < G$ if H is an option of G. When proving a fact about games G, we will assume, by induction, that the fact holds true for all options of G. Usually, for the trivial game with no options, the fact will vacuously hold, and no explicit base case will be required!

A.2 Examples

Example A.1. Show that the sum of the first n positive integers is $n(n+1)/2$.

Proof:
$$1 + 2 + \cdots + n = (1 + 2 + \cdots + (n-1)) + n \tag{A.1}$$
$$= \frac{(n-1)n}{2} + n \quad \text{(by induction)} \tag{A.2}$$
$$= \frac{n(n+1)}{2}. \tag{A.3}$$

The base case is $n = 0$, and $n(n+1)/2 = 0$, as it should. $\qquad\square$

Note that the base case of $n = 0$ is the *only* case not handled by (A.1). In particular, when $n = 0$, there are no terms in $1 + 2 + \cdots + n$, and so there is no n to separate out.[2]

Exercise A.2. Prove that the sum of the first n *odd* positive integers is n^2.

---

[1]Such orders are called *well quasi-orders* and logicians sometimes call this technique ϵ-induction. See Problem 1 for more formalism.

[2]This example is presented *only* because it is almost always the first example of a proof by induction and so will be familiar! A *much better* argument in this case is to use Gauss's trick — note that the average value of a summand is $(n+1)/2$ and hence the sum must be $n(n+1)/2$.

Example A.3. (Two-heap NIM**)** The game of two-heap NIM is played with two heaps of counters of sizes m and n. A move consists of removing any number of counters from either heap. The person who takes the last counter wins. Prove that the player on move can guarantee a win *if and only if* $m \neq n$.

Proof: Denote the position by the ordered pair (m, n), where without loss of generality $m \leq n$. If $m < n$, the player on move can move to (m, m), which, by induction, she wins. If, on the other hand, $m = n$, every move leaves heaps of differing sizes, which, by induction, allows the opponent to guarantee a win; since all moves lose, the position is losing. □

Note that no base case is required in the last theorem. When, for example, $m = n = 0$, the last sentence still holds: there is no legal move, and so every move really does leave heaps of differing sizes.

Exercise A.4. The following variant of NIM is played with a single heap starting with n counters. A move consists of removing between one and ten counters, and the player who takes the last counter wins.

(a) Prove that the second player wins *if and only if* $n \equiv 0 \pmod{11}$.

(b) Explain why your proof requires no base case.

When trying to solve any problem, first try to break up your problem into smaller pieces. In (A.1), there was one very natural way to do this, and induction followed naturally. The next two examples illustrate this even more clearly.

Example A.5. Show that the sum of the interior angles of an n-sided polygon is $180 \cdot (n - 2)$.

Proof: Let an n-sided polygon be given. Find an interior diagonal of this polygon; that is, a line connecting two of its vertices that passes through the interior of the polygon and meets it only at its endpoints.[3]

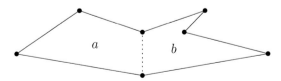

---

[3]Showing that a diagonal exists is surprisingly difficult! Choose three consecutive vertices A, B, and C such that the interior angle $\angle ABC$ is less than 180 degrees. If AC is a diagonal then we are done. Otherwise, there are points of the polygon on the boundary or interior of the triangle ABC. Among these choose a vertex of the polygon D such that the line through D parallel to AC is as close to B as possible. Then, BD is a diagonal.

The sum of the angles of the original polygon is equal to the sum of the angles of the two parts of the polygon formed by the diagonal. Let these parts have a and b sides, respectively. The sum $a + b$ is equal to $n + 2$ since each original side of the polygon is a side of one of the parts and the diagonal counted twice as a side of both parts. By induction, the sum of the angles of the polygon is

$$180 \cdot (a - 2) + 180 \cdot (b - 2) = 180 \cdot (a + b - 4) = 180 \cdot (n - 2) \text{ degrees.}$$

The base case of the induction occurs when we cannot find such a diagonal; that is, if $n = 3$, but the result is known in that case.   □

Example A.6. Let the Fibonacci numbers a_n be given by

$$a_n = \begin{cases} 0 & \text{if } n = 0, \\ 1 & \text{if } n = 1, \\ a_{n-1} + a_{n-2} & \text{if } n \geq 2. \end{cases}$$

Show that $a_n = a_r a_{n-r-1} + a_{r+1} a_{n-r}$ for $0 \leq r < n$.

Proof:
$$a_n = a_{n-1} + a_{n-2} \tag{A.4}$$
$$= (a_r a_{n-r-2} + a_{r+1} a_{n-r-1}) + (a_r a_{n-r-3} + a_{r+1} a_{n-r-2}) \tag{A.5}$$
$$= a_r (a_{n-r-2} + a_{n-r-3}) + a_{r+1} (a_{n-r-1} + a_{n-r-2}) \tag{A.6}$$
$$= a_r a_{n-r-1} + a_{r+1} a_{n-r}. \tag{A.7}$$

Here, Equation (A.5) is obtained by induction and Equation (A.7) by the definition of the Fibonacci numbers. Now, to figure out the base cases, we need to see how the argument above can fail. Equation (A.4) fails when $n = 0$ or $n = 1$. Equation (A.5) fails when $r = n - 1$ or $r = n - 2$, since the statement of the theorem insists that $r < n$. In other words, when invoking the induction hypothesis, we need $r < n - 1$ to expand a_{n-1} and $r < n - 2$ to expand a_{n-2}. Equations (A.6) and (A.7) introduce no further problematic cases. So, the base cases are the following:

- $n = 0$: No $0 \leq r < n$ exists, so the theorem is vacuously true.

- $n = 1$: Then $r = 0$ and $a_0 a_0 + a_1 a_1 = 0 + 1 = 1 = a_1$.

- $r = n - 1$: $a_{n-1} a_0 + a_n a_1 = a_n$.

- $r = n - 2$: $a_{n-2} a_1 + a_{n-1} a_2 = a_{n-2} + a_{n-1} = a_n$.   □

Note that there is really no need to differentiate between "base cases" of an induction and "special cases" of any proof. One could rewrite the whole proof without any "base case," and the choice is a matter of style:

Sketch of alternate proof: If $n = 0$ or $n = 1$, we can verify that the theorem holds. Otherwise,

$a_n = a_{n-1} + a_{n-2}, \ n \geq 2$

$$= \begin{cases} a_{n-1}a_0 + a_n a_1 = a_n & \text{if } r = n-1, \\ a_{n-2}a_1 + a_{n-1}a_2 = a_{n-2} + a_{n-1} = a_n & \text{if } r = n-2, \\ (a_r a_{n-r-2} + a_{r+1}a_{n-r-1}) + & \\ \quad (a_r a_{n-r-3} + a_{r+1}a_{n-r-2}) & \\ \quad = a_r(a_{n-r-2} + a_{n-r-3}) + & \text{otherwise (by induction).} \\ \quad a_{r+1}(a_{n-r-1} + a_{n-r-2}) & \\ \quad = a_r a_{n-r-1} + a_{r+1}a_{n-r} & \end{cases}$$

$\square$

There is another proof, which is arguably even better — because it leads to a deeper understanding of *why* this identity holds and also to methods for generating similar identities. Consider the matrix

$$A = \begin{pmatrix} 0 & 1 \\ 1 & 1 \end{pmatrix} = \begin{pmatrix} a_0 & a_1 \\ a_1 & a_2 \end{pmatrix}.$$

Then, it is easy to check (by induction!) that

$$A^n = \begin{pmatrix} a_{n-1} & a_n \\ a_n & a_{n+1} \end{pmatrix}.$$

The desired identity is then just a consequence of $A^n = A^r A^{n-r}$.

Theorem A.7. (Euler's Formula) *For any planar embedding of a graph with E edges, V vertices, C connected components, and R regions, we have*

$$E - V - R + C + 1 = 0.$$

Proof: Fix a planar embedding as above. Removing an edge either merges two regions into one or increases the number of components by 1. So, by induction, either

$$(E - 1) - V - (R - 1) + C + 1 = 0 \quad \text{or}$$
$$(E - 1) - V - R + (C + 1) + 1 = 0.$$

It trivially follows that $E - V - R + C + 1 = 0$. For the base case, if $E = 0$, then $V = C$ and $R = 1$, and Euler's Formula again holds. $\square$

A.3 Why Is Top-Down Induction Better?

The usual axiomatic presentation of mathematical induction suggests that to prove some statement $P(n)$ for $n \geq 0$,

1. prove the base case, $P(0)$;

2. then, for the induction step, assume that $P(n)$ and prove $P(n+1)$.

There are many reasons why top-down induction is better to use than the axiomatic approach. The most important is that in complicated proofs using axiomatic induction it is easy to overlook some cases of size $n + 1$. This is a popular pitfall among even the best of mathematicians! One such false proof will exemplify the pitfall.

Example A.8. Prove that n internal points triangulate a triangle into $2n + 1$ smaller triangles. In other words, n points are placed in the interior of a triangle indicated by three points. If non-crossing line segments (*edges*) are introduced connecting the $n + 3$ points until no more can be added, then the triangle will contain $2n + 1$ regions:

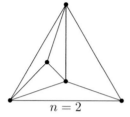
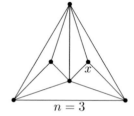

$$n = 2 \qquad\qquad n = 3$$

False proof: For a base case, if there are $n = 0$ added points, then there are $2n + 1 = 1$ triangles. For the inductive step, assume that if there are n points, there must be $2n+1$ triangles. To prove the case for $n+1$, add one more point, say x, to the triangle. This removes one region but creates three, leaving a total of $2n + 1 - 1 + 3 = 2(n+1) + 1$ regions.   □

This argument is very seductive; in fact, most are taken in by it unless warned it is fallacious. The reason the proof is incomplete is that we have not argued about *all* possible triangulations — only those that can be arrived at from smaller triangulations. You may be tempted to patch the proof by permitting the addition of points along previous edges, but that still would not be good enough! With some work, you can come up with a triangulation on $n = 11$ interior points where each point is incident to at least five edges. This triangulation could not have been arrived at from a smaller triangulation with the addition of one point — the last point added would only be incident to three edges (or four edges in the proposed "patch").

Correct proof of Example A.8: Begin with a triangulation on n interior points. Remove one of the points. If the point had k incident edges, then k triangles will be removed, leaving a k-sided polygonal region. A k-sided polygon can be divided into $k - 2$ triangles (by an argument parallel to Example A.5). By induction, this new triangulation has $2n - 1$ triangles, but it has $k - (k - 2)$ or two fewer triangles than the original. So, the original triangulation had $2n + 1$ triangles. The base case of $n = 0$ is trivial.   □

Being able to write consistently correct proofs is the most important advantage with top-down induction. But, a few more advantages are worth noting:

- The axiomatic model's assumption that you are always trying to prove a statement $P(n)$ for $n \geq 0$ is simplistic. First, in actual proofs, the parameter n might take on many different forms. Further, determining what in your proof *should* serve the role of n is not always obvious. Of course, those investigating the fundamentals of mathematics would consider this simplistic view a benefit, but most students are interested in understanding how to apply induction in the widest possible context.

- In a complicated proof, it is unclear what is the *right* base case to choose, particularly if the induction step has not been proved yet. In Example A.6, it is hard to foresee what base case is needed before doing the induction argument; so why do the base case first? Further, top-down induction gives guidelines for how to choose the strongest possible, minimal-sized base case.

- Students often wonder, "How do I know when to use induction?" This is a surprisingly hard question to answer. The top-down induction paradigm sidesteps this issue somewhat, since there is no need for the prover to know in advance that induction is required. Through the process of breaking up a problem or expanding or massaging an expression, if a smaller case ever appears, induction can be used freely.

- Induction on several variables is completely demystified. The prover need not know in advance on what parameters induction will be used, nor whether induction will be used on several variables.

- The axiomatic induction suggests that to prove $S(n)$, first assume (by way of induction) $S(n)$. This *sounds* circular. Of course it is not, since you would then proceed to prove $S(n + 1)$. But since you are trying to prove $S(n)$, doesn't it make more sense to assume $S(n - 1)$ and prove $S(n)$? Top-down induction, like strong induction, goes one step further, allowing you to assume $S(m)$ for $m < n$.

A.4   Strengthening the Induction Hypothesis

Strengthening the induction hypothesis is a technique of trying to prove more than you need to. It may seem counterintuitive that this could make life easier;

but if you try to prove more, you also get to *assume* more (by induction). Here is an example:

Example A.9. Define the Fibonacci numbers a_n as in Example A.6. Suppose that $n = 2k + 1$ is odd. Prove that $a_n = a_{k+1}^2 + a_k^2$.

The natural first step is to write

$$a_{2k+1} = a_{2k} + a_{2k-1}$$
$$= a_{2k} + a_k^2 + a_{k-1}^2 \quad \text{(by induction)}.$$

It is hard to go further, since we have no formula for a_n when n is even. Let's be optimistic and try to see what a_{2k} would need to complete the proof. Well, if a_{2k} were equal to $a_{k+1}^2 - a_{k-1}^2$, that would do it easily! This suggests *strengthening the induction hypothesis*. So, we will aim to prove a statement that deals with both cases. Rather than using the form above for a_{2k}, experience suggests that something more similar to the form for a_{2k+1} might make the algebra simpler. Such a form is easily found:

$$a_{k+1}^2 - a_{k-1}^2 = (a_{k+1} - a_{k-1})(a_{k+1} + a_{k-1})$$
$$= a_k(a_{k+1} + a_{k-1})$$
$$= a_k a_{k+1} + a_k a_{k-1}.$$

So, we need to prove that

$$a_{2k+1} = a_{k+1}^2 + a_k^2 \quad \text{and}$$
$$a_{2k} = a_k a_{k+1} + a_k a_{k-1}.$$

Proof:
$$a_{2k+1} = a_{2k} + a_{2k-1}$$
$$= a_k a_{k+1} + a_k a_{k-1} + a_k^2 + a_{k-1}^2 \quad \text{(by induction)}$$
$$= a_k a_{k+1} + (a_k + a_{k-1})a_{k-1} + a_k^2$$
$$= a_{k+1}(a_k + a_{k-1}) + a_k^2$$
$$= a_{k+1}^2 + a_k^2.$$

$$a_{2k} = a_{2k-1} + a_{2k-2}$$
$$= a_k^2 + a_{k-1}^2 + a_{k-1}a_k + a_{k-1}a_{k-2} \quad \text{(by induction)}$$
$$= a_k(a_k + a_{k-1}) + a_{k-1}(a_{k-1} + a_{k-2})$$
$$= a_k a_{k+1} + a_k a_{k-1}.$$

We omit the base case. $\square$

A.5   Inductive Reasoning

Example A.10. (Pick's Theorem) The *lattice points* in the plane are the points with integer coordinates. Let P be a polygon that does not cross itself (i.e., a *simple* polygon) such that all of its vertices are lattice points. Let p be the number of lattice points that are on the boundary of the polygon (including its vertices), and let q be the number of lattice points that are inside. Prove that the area of the polygon is $\frac{p}{2} + q - 1$.

The key to the proof of Pick's Theorem is that, when using induction, we can assume that the statement we are trying to prove is true for *simpler* polygons rather than *smaller* polygons. As in Example A.5, we can divide the polygon into smaller polygons and use induction. But, proving Pick's Theorem for a triangle still looks challenging. It's not hard to prove Pick's Theorem for right triangles (and rectangles) aligned with the x- and y-axes, though, since it is easier to count lattice points on the boundary and interior of these right triangles. So, how do we reduce the problem to just right triangles? Read on ....

The following lemma is a simple exercise.

Lemma A.11. *Let P_1 and P_2 be lattice polygons whose union, P, is also a simple lattice polygon. Assume that P_1 and P_2 intersect only along a set of edges (possibly only one). Then, the following hold:*

Addition Lemma: If P_1 and P_2 satisfy Pick's Theorem, then so does P.

Subtraction Lemma: If P and P_1 satisfy Pick's Theorem, then so does P_2.

Halving Lemma: If P_1 and P_2 are isomorphic, then P satisfies Pick's Theorem if and only if P_1 does.

The areas of P_1 and P_2 add to P. So, to prove the lemma, it suffices to show that Pick's Theorem is also additive; i.e., $(\frac{p_1}{2} + q_1 - 1) + (\frac{p_2}{2} + q_2 - 1) = \frac{p}{2} + q - 1$.

Now for a picture proof of Pick's Theorem:

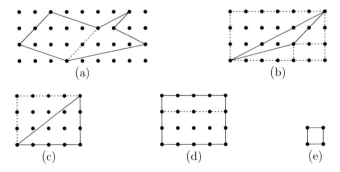

(a)   (b)

(c)   (d)   (e)

A polygon (a) with at least four sides can be separated into two regions, each which, by induction, satisfy Pick's Theorem, and so by the Addition Lemma, the original polygon satisfies Pick's Theorem as well.

A triangle (b) can be formed by the difference between a large rectangle, and smaller rectangles and right triangles all aligned with the x- and y-axes. Each of the rectangles and right triangles satisfy Pick's Theorem (by induction), and thus by the Subtraction Lemma, so does the arbitrary triangle.

For a right triangle (c), adjoin another right triangle to form a rectangle. The rectangle satisfies Pick's Theorem by induction, and thus by the Halving Lemma, so does the right triangle.

A rectangle (d) can be divided into two rectangles by a horizontal or vertical line; apply induction.

Finally, for the base case of a unit square, Pick's Theorem yields $(\frac{p}{2}+q-1) = \frac{4}{2} + 0 - 1 = 1$, which is indeed the unit square's area.

Problems

1. Formally, the principle of top-down induction can be stated as follows. Let universe U be a partially ordered set with the property that every non-empty subset of U has at least one minimal element. If

$$\forall n : [\forall m < n : P(m)] \ \Rightarrow \ P(n),$$

   then
$$\forall n : P(n).$$

   (Although not relevant to this problem, note that if n is minimal, $\forall m < n : P(m)$ is vacuously true. This is usually called a *base case*.)

   Prove the principle of top-down induction by contradiction. (*Hint:* Let $T = \{n : P(n) \text{ is false}\}$. You will prove $T = \emptyset$.)

2. To highlight the importance of the assumption "every non-empty subset of U has at least one minimal element," show that the statement $x = 0$ about real numbers $x \in [0, 1]$ satisfies the inductive condition above but is *not* true of every real number in $[0, 1]$. In other words,

   (a) Sketch a "good-looking" top-down induction proof that $x = 0$ for all $0 \leq x \leq 1$.

   (b) Explain how every subset of reals in $[0, 1]$ may not have a minimal element by giving a specific counterexample.

3. Prove that any triangulation of an n-sided polygon yields $n - 2$ triangles. This should complete the correct proof to Example A.8.

4. Complete the proof of Pick's Theorem by proving Lemma A.11.

5. Prove that any rational number between 0 and 1 can be written as a sum of distinct fractions with numerator 1. As an example, $\frac{7}{11} = \frac{1}{2} + \frac{1}{11} + \frac{1}{22}$.

6. Prove the *Arithmetic-Mean-Geometric-Mean Inequality for n Numbers*. For any n non-negative numbers $a_1, a_2, \ldots, a_n$,

$$\frac{a_1 + a_2 + \cdots + a_n}{n} \geq \sqrt[n]{a_1 a_2 \cdots a_n}.$$

Let $P(n)$ be the theorem for fixed n.

(a) For a warm-up, prove $P(1)$ and $P(2)$.

(b) If $n = 2m$, prove that $P(m)$ implies $P(n)$.

(c) Prove that $P(n+1)$ implies $P(n)$ (this is *backward induction*).

(d) Explain how using induction you have proved the theorem for all n.

Preparation for Appendix B

To the instructor: We highly encourage anyone learning combinatorial game theory to do so hand-in-hand with CGSuite. Doing calculations by hand builds intuition, confirming the calculations using CGSuite builds confidence, and delegating long calculations to CGSuite builds speed.

Appendix B
CGSuite

Besides black art, there is only automation
and mechanization.

Federico García Lorca

The Combinatorial Game Suite (CGSuite) [Sie03] is an open-source program authored by Aaron Siegel that does all of the algebraic manipulations of games. Written in Java, the program is platform independent. Its features include:

- a great user interface;

- the ability to manipulate short games and loopy games;

- lots of built-in games;

- a graphical explorer to navigate a position's game tree;

- a programming language, CGScript, for automating calculations and implementing rules for new games.

The goal of this appendix is to give you a brief introduction to the software. Much of this appendix was excerpted, with permission, from the CGSuite tutorial at www.cgsuite.org.

B.1 Installing CGSuite

Installing the software on most modern systems is easy. Go to www.cgsuite.org and follow the directions. Just download and unpack the software, and you should be ready to go. Should anything go wrong in the installation process, double-check that you have a sufficiently recent version of Java.

B.2   Worksheet Basics

The most fundamental part of CGSuite is the worksheet, where you can type commands and perform calculations directly. If you are new to CGSuite, you should launch CGSuite and begin experimenting by typing commands into the worksheet.

Entering games

You can type games directly into the worksheet. For example, click on the window labeled "Worksheet 1" and enter the following:

    1/2 + vv*2

This represents the game $\frac{1}{2}+\Downarrow+*2$. In general, you can enter combinations of numbers, ups, downs, and nimbers just as you would expect, but sometimes you'll need an explicit + to add games. The symbol ^ is used for ↑, v for ↓, and * for nimbers. To enter, say, ⇑*, you could type either ^^^^ or ^4. More complicated games can be entered using braces and slashes. For example:

    {2||1|0,*}

Expressions containing slashes must be enclosed in braces. That is, to enter the game 1|0, you must type {1|0}. Ambiguous expressions, such as {1|0|-1}, will be rejected; you would need to enter {1||0|-1} (or {1|{0|-1}}) or {1|0||-1} (or {{1|0}|-1}) instead.

CGSuite recognizes a wide variety of common games and displays them using standard shorthand notation (such as +₂ for tiny-two). Here are a few to try (make sure to enter each on a separate line):

    {0||0|-2}
    {0|v*}
    {1,1+*|-1,-1+*}

The game {0||0|-2} could also be entered as 2.Tiny. The use of the period in the notation 2.Tiny is a consequence of everything in CGSuite (including integers like 2) being an *object,* and we are saying "apply the operation Tiny to the object 2." Once you get used to the use of the period, you'll be good to go.

Operations on games

There are nine comparison operators:

    ==   <=   >=   <   >   !=   <| |>   <>

(From left to right: equal, less than or equal, greater than or equal, less than, greater than, not equal, less than or confused with, greater than or confused with, and confused with.) For example, try entering:

```
^^ > *
0 <| ^*
```

Other common operations can be entered as method calls. Here are some examples to try:

```
{3||2|1}.Mean
{3||2|1}.Temperature
{3||2|1}.Thermograph.Plot()
{^^|vv}.AtomicWeight
{3||2|1}.Cool(1/2)
{3||2|1}.Freeze
*.Heat(3)
*.Overheat(1,2)
*7.OrdinalSum(1)
```

Many more are available as well; see the glossary of methods in CGSuite's tutorial for a complete list. Variable assignments are permitted, e.g.:

```
G := {3||2|1}
G.Freeze
```

Multiple statements can be strung together as a single command using semicolons, so the above could be rewritten on a single line as:

```
G := {3||2|1}; G.Freeze
```

Note that output is generated only for the last command. If a command ends with a semicolon, no output will be generated at all.

Types and canonical forms

Every object in CGSuite has a type associated with it. Objects such as ^^, {3||2|1}, etc. are canonical games, but CGSuite also supports games that are not in canonical form. In this way, positions in a game such as DOMINEERING are distinguished from their canonical forms.

For example, try entering the following:

```
H := game.grid.Domineering(Grid(4,4))
```

This assigns to H the DOMINEERING position represented by an empty 4×4 grid. Note that CGSuite does not yet try to calculate the canonical form of H. The canonical form of H can be obtained by explicit user request, using the `CanonicalForm` method:

H.CanonicalForm

Several other games are included as examples, including AMAZONS, CLOB-BER, FOX & GEESE, KONANE, and TOADS & FROGS. Typically, positions can be entered as a sequence of strings, separated by commas; each string corresponds to a single row in the grid. For example, try entering:

```
A := game.grid.Amazons("x..#.|o....")
```

As always, `A.CanonicalForm` calculates the canonical form of A.

Exercise B.1. Let G be the game $\{2 \parallel 1 \mid *\}$. Calculate the canonical forms of $G + G$ and $G + G + G + G$. Compute the mean and temperature of G. How does G compare with 1 and $1*$?

Exercise B.2. Use the `LeftOptions` and `RightOptions` commands to determine the canonical form of $\Uparrow$.

Exercise B.3. Let H be the canonical form of the 4×4 DOMINEERING rectangle that we calculated above. Try comparing H with small positive and negative numbers. Make a conjecture as to whether or not H is an infinitesimal, and then use `H.IsInfinitesimal` to test your conjecture. How does H compare with various tinies?

Exercise B.4. Try to find constraints on positive numbers a, b, c ($a > b$) such that $\{a \mid b\} + \mathord{+}_c = \{a \mid_c \mid b \mid_c\}$.

Graphically exploring positions

An intuitive interface for exploring game positions is provided by the graphical explorer. After reviewing the CGSuite tutorial's section entitled "Using the Explorer," try these exercises.

Exercise B.5. Find the best moves from a 4×4 DOMINEERING rectangle. Then, find the "sensible lines of play," observing that one move for each player reverses out.

Exercise B.6. Find the sensible lines of play for the following 3×4 CLOBBER position:

Then, calculate its atomic weight. If you have the patience, find the sensible lines of play for each of its sensible options, and observe that CLOBBER is not an easy game!

Exercise B.7. Calculate the canonical form and atomic weight of the following 3×7 CLOBBER position from [AGNW05]:

Its canonical form is surprisingly complicated given that Right has just one piece.

B.3 Programming in CGSuite's Language

You can write your own programs in the worksheet in CGSuite's built-in language, CGScript, a language similar to Maple or Visual Basic.

Work through the full in-application tutorial, and you will be prepared to implement your own games.

Appendix C

Solutions to Exercises

Solutions for Chapter 1

1.7: (page 15)

(a) If m, n are both even, then the second player should play a symmetry strategy — play the same move but rotate the board by 180 degrees.

(b) If one of m, n is even and the other odd, then the first player should take the central two squares and then play a symmetry strategy — play the same move but rotate the board by 180 degrees.

1.9: (page 16) In this game, the number of moves in the game is fixed! Each move creates one more piece, at the start of the game there is one piece, and at the end there are 30. So the game lasts 29 moves, and the first player wins no matter how she plays.

1.17: (page 23) The middle move is the only non-loony move.

1.18: (page 24)

(a) Both positions are loony since the next player can take the fourth and fifth coins in the top row and play first in the rest of the game — or can cut the string joining those coins and move second in the rest of the game.

(b) Alice should double-deal in the left-hand position because then Bob has to move first in the long chains. In the other position, Alice should take both coins and then cut the lower-right string.

(c) Left-hand game: Alice 6, Bob 4; right-hand game: Alice 8, Bob 2.

1.20: (page 27) Alice 4, Bob 5.

Solutions for Chapter 2

2.2: (page 36) If $a = b$ the outcome type is $\mathcal{P}$, while if $a \neq b$ the outcome type is $\mathcal{N}$. Since the game is impartial, it's clear that types $\mathcal{L}$ and $\mathcal{R}$ can't arise (any winning strategy for Left as first player is equally a winning strategy for Right). To justify the rest of the claim, note that when the first player moves from a position with $a = b$ it leaves a position with two heaps of unequal sizes, and the second player can utilize the Tweedledum-Tweedledee strategy (i.e., restore the equality of the heaps). On the other hand, if the two heaps are initially of different sizes, the first player can take enough counters from the larger one to make them equal and then win the resulting position in her new role as second player.

2.3: (page 36) The position with no unoccupied cells is of type $\mathcal{P}$ (since whoever's turn it is to move has lost). In a position of type $\mathcal{N}$, both players must have a move available (since they have winning strategies as the first player) and since there is no two-cell position with this characteristic, the smallest such position has three cells (any 2×2 square with one cell removed will do). For type $\mathcal{L}$ the Left player must have a move, so at least two cells are required and ⊟ shows that two cells suffice. Similarly for type $\mathcal{R}$ we can use ⊡.

2.8: (page 40)

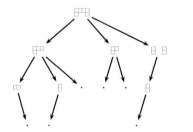
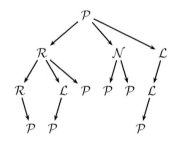

2.10: (page 41) The only difference is that from the top square Left can move to a $\mathcal{P}$-position, and so the top square has outcome class $\mathcal{L}$.

2.14: (page 43) Fix an impartial game G. By induction, all options are either in $\mathcal{N}$ or in $\mathcal{P}$, and since the left and right options are identical, some $G^R \in \mathcal{P}$ if and *only if* some $G^L \in \mathcal{P}$. Hence, $G \in \mathcal{P}$ or $G \in \mathcal{N}$. (In other words, we can only be in the upper-left or lower-right corner of the table shown in Observation 2.6.)

2.16: (page 43) The terminal positions have no options, and so the statement "every option is in B" is vacuously satisfied. So, terminal positions must be assigned to set A.

2.18: (page 44) If G_1 is a $\mathcal{P}$-position, then G is an $\mathcal{N}$-position since G_1 is an option of G. If G_1 is an $\mathcal{N}$-position, then the only option of G_2 is an $\mathcal{N}$-position. Hence $o(G_2) = \mathcal{P}$ and $o(G) = \mathcal{N}$.

2.20: (page 45) Doing all arithmetic modulo 7, the $\mathcal{P}$-positions are exactly those where $n \equiv 0$ or $n \equiv 2$. From such a position, all moves are in $\{0-1, 0-3, 0-4, 2-1, 2-3, 2-4\}$, which are equivalent to $\{6, 4, 3, 1, 6, 5\}$ and are all $\mathcal{N}$-positions. On the other hand, if $n \in \{1, 3, 4, 5, 6\}$, there are winning moves to $\{0, 0, 0, 2, 2\}$, respectively.

2.21: (page 46)

(a) If Left adopts the strategy of always moving the roller to reduce just one bump, then she has three moves available while Right has at most two, so she will win moving first or second. Thus, the outcome type is $\mathcal{L}$.

(b) If Left moves first she can leave $[1, \circ, 0, 5]$. Now Right's move is forced to $[1, 0, 4, \circ]$. Next, if Left always moves the roller one position, she will eventually win: Right has the final move on the rightmost bump, but Left has one move remaining. Left's strategy if Right moves first is similar, and the outcome type is $\mathcal{L}$.

(c) Each player can respond to the other's move(s) by clearing away all the height 1 bumps that have just been created. So, the second player always wins, i.e., the game is of type $\mathcal{P}$.

(d) As in the first example, if either player has a numerical advantage (i.e., for Left, more 1s to the left of the roller than to the right), then they can win. If there are the same number of 1s to the left and right of the roller, then the previous player wins since any move the first player makes leaves their opponent with a numerical advantage.

2.26: (page 49) The position ⫿⫿⫿ has poison pairs at both ends and reduces to $\cdots$ ⫿ $\cdots$. So, it is of type $\mathcal{N}$ and we win by knocking over either of the two middle dominoes. Eliminating poison pairs from ⫿⫿⫿⫿⫿ leaves $\cdots$ ⫿ $\cdots$ ⫿ $\cdots$ ⫿ and two more rounds of reduction leaves the empty string, so this position is of type $\mathcal{P}$.

2.27: (page 50) It's a matter of opinion, of course! We find the second argument very attractive and elegant, but if it was the only one we knew we'd be left with a vague feeling that we didn't know *why* these were the $\mathcal{P}$-positions. On the other hand, the argument that starts with "poisoned pairs" is a bit more hands on (and perhaps messy) but gives a really concrete understanding of exactly why the result holds. We prefer to know them both!

2.28: (page 51)

a	(a) $a22$	(b) $a23$
0	$\mathcal{P}$	$\mathcal{R}$
1	$\mathcal{N}$	$\mathcal{R}$
2	$\mathcal{N}$	$\mathcal{R}$
3	$\mathcal{L}$	$\mathcal{P}$
4	$\mathcal{L}$	$\mathcal{L}$
5	$\mathcal{L}$	$\mathcal{L}$
⋮	$\mathcal{L}$	$\mathcal{L}$

2.29: (page 51) If Left can win by taking a, then there is nothing to prove. Among other things this covers the case when $\mathbf{w}b$ is empty. If Left has a winning move to $a'\mathbf{w}b$, $0 < a' < a - 1$, after Right's move to $a'\mathbf{w}b'$, then Left has a winning move to $a''\mathbf{w}b'$. However, Left has the same winning move available from $(a - 1)\mathbf{w}b'$.

The point is that the options from leftmost-heap size a', where $0 < a' < a$, form a subset of those from $a - 1$. So, the move to $a - 1$ is at least as good as the move to a', for it burns fewer bridges.

2.31: (page 52) If $a > L(\mathbf{w}b)$, then Left can win moving first by decreasing pile a to exactly $L(\mathbf{w}b)$, from which she wins moving second by the definition of $L(\cdot)$. If, however, $a \leq L(\mathbf{w}b)$, then every move by Left loses, for it leaves a position of the form $a'\mathbf{w}b$ for $0 \leq a' < L(\mathbf{w}b)$. Similarly, Right wins moving first *if and only if* $b > R(a\mathbf{w})$. The observation follows.

Solutions for Chapter 3

3.6: (page 66) Note that $G + H + \boxed{}$ is in $\mathcal{N}$. For both players, the winning first move is in H.

Solutions for Chapter 4

Prep 4.1: (page 74) Left simply responds locally. When playing $G_1 + G_2$, if Right moves on G_1, Left makes a response on G_1 that wins moving second on it. Similarly, Left responds to moves in G_2 by moving on G_2. Eventually, Right will run out of moves in both G_1 and G_2 and so will lose $G_1 + G_2$.

Prep 4.2: (page 74) One example is $G_1 = G_2 = \{0 \mid -1\}$.

4.6: (page 79)

(1) By definition, $G + H = \{\mathcal{G}^L + H, G + \mathcal{H}^L \mid \mathcal{G}^R + H, G + \mathcal{H}^R\}$, and by induction, all the simpler games are commutative; for example, $G^L + H = H + G^L$ for $G^L \in \mathcal{G}^L$. So,

$$
\begin{aligned}
G + H &= \{\mathcal{G}^L + H, G + \mathcal{H}^L \mid \mathcal{G}^R + H, G + \mathcal{H}^R\} \\
&= \{H + \mathcal{G}^L, \mathcal{H}^L + G \mid H + \mathcal{G}^R, \mathcal{H}^R + G\} \quad \text{(by induction)} \\
&= H + G.
\end{aligned}
$$

(2) To save page width, we will focus on the left options only:

$$
\begin{aligned}
&[(G + H) + J]^L \\
&= \{(G + H)^L + J, (G + H) + \mathcal{J}^L\} \\
&= \{(\mathcal{G}^L + H) + J, (G + \mathcal{H}^L) + J, (G + H) + \mathcal{J}^L\} \\
&= \{\mathcal{G}^L + (H + J), G + (\mathcal{H}^L + J), G + (H + \mathcal{J}^L)\} \text{ (by induction)} \\
&= [G + (H + J)]^L.
\end{aligned}
$$

Since the argument proceeds similarly for the right options, $(G+H)+J = G + (H + J)$.

4.7: (page 80)

4.8: (page 80) All the statements follow from the following observation: Left has a winning move in G if and only if Right has a winning move in $-G$.

4.9: (page 80)

$$
\begin{aligned}
-(-G) &= -\{-\mathcal{G}^R \mid -\mathcal{G}^L\} \\
&= \{-(-\mathcal{G}^L) \mid -(-\mathcal{G}^R)\} \\
&= \{\mathcal{G}^L \mid \mathcal{G}^R\}.
\end{aligned}
$$

4.10: (page 80)

$$-(G + H) = -\{\mathcal{G}^L + H, G + \mathcal{H}^L \mid \mathcal{G}^R + H, G + \mathcal{H}^R\}$$
$$= \{-\mathcal{G}^R - H, -G - \mathcal{H}^R \mid -\mathcal{G}^L - H, -G - \mathcal{H}^L\}$$
$$= \{(-\mathcal{G})^L + (-H), (-G) + (-\mathcal{H})^L \mid (-\mathcal{G})^R + (-H), (-G) + (-\mathcal{H})^R\}$$
$$= (-G) + (-H).$$

4.11: (page 80) Reflexivity: clearly, $o(G + X) = o(G + X)$ for any game X.

Symmetry: for any game X, if $o(G + X) = o(H + X)$ then $o(H + X) = o(G + X)$

Transitivity: if, for any game X, $o(G + X) = o(H + X)$ and $o(H + X) = o(K + X)$, then $o(G + X) = o(K + X)$.

4.14: (page 81) If Left moves northwest and fires upward, then Right moves up and fires to the right: Left has at most three moves remaining whereas Right has four. If Left moves right and fires southeast, Right moves up and blocks Left. If Left moves southwest and fires anywhere but the square just vacated, then Right occupies this square and fires anywhere but northwest and Right has more moves remaining.

If Right makes any move but the one shown, Left moves right and fires back. Left has at least as many moves as Right. (*Note:* Left may have better moves, but this wins.)

4.17: (page 82)

$$
\begin{array}{rll}
G & = & G + 0 \qquad\qquad \text{(Theorem 4.4)} \\
  & = & G + (J - J) \quad \text{(Corollary 4.15)} \\
  & = & (G + J) - J \quad \text{(Theorem 4.5)} \\
  & = & (H + J) - J \quad \text{(Assumed)} \\
  & = & H + (J - J) \quad \text{(Theorem 4.5)} \\
  & = & H + 0 \qquad\qquad \text{(Corollary 4.15)} \\
  & = & H \qquad\qquad\quad \text{(Theorem 4.4)}
\end{array}
$$

4.18: (page 82) By two applications of Theorem 4.16 (and, as always, commutativity of $+$),

$$G + H = G' + H = G' + H'.$$

4.20: (page 83) If $G \geq H$ then, for any X, $o(G + X) \geq o(H + X)$, which is the same as $o(H + X) \leq o(G + X)$, and therefore $H \leq G$.

Now, $G = H$ *if and only if* $o(G + X) = o(H + X)$, in particular $o(G + X) \geq o(H + X)$ and $o(G + X) \leq o(H + X)$; that is, $G \geq H$ and $G \leq H$.

4.22: (page 84) $2 \Rightarrow 4$: Suppose that Left can win moving second on G and can win moving first on X. To win on $G + X$ playing first, she plays a winning move in X (treating X as an isolated game) and then responds locally as in $2 \Rightarrow 3$.

4.29: (page 87)

1. $= \{0 \mid 1\} = \frac{1}{2}$, which is in $\mathcal{L}$.

2. $= \{0 \mid 0\} = *$ and so $= \{0, *, -1 \mid 0, *, 1\} = \{0, * \mid 0, *\}$, which happens to be $*2$ (not yet defined) and is not born until day 2.

3–4. The MAIZE position is a $\mathcal{P}$-position with value 0. In MAZE, we get an $\mathcal{N}$-position; its value is $*2$, born on day 2.

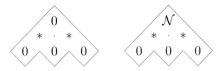

5–6. The MAIZE position is in $\mathcal{P}$ and has value 0. The MAZE position is in $\mathcal{L}$ and has value $\frac{1}{2}$, which is born on day 2.

4.32: (page 88)

(a) Left moving first moves to

Moving second, play is forced, and Left again gets the last move to

(b) The second player wins the difference game

In particular, if one player plays in one component, the second player plays in the other (with Right preferring over). Diagrammatically,

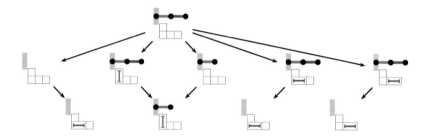

4.35: (page 93) If Left or Right play any of the ticked options, then the opponent plays the same (but negative) version (e.g., to $B - B$), reducing the game to 0.

The only first move for Left not covered is her move to $A - G'$. Now Right wins by moving to $A^R - G'$. To see this, note that if Left moves to $X - G'$ then Right plays to $X - X = 0$ and wins. If Left moves to $A^R - H$, then this is an option from $A^R - G$. However, $A^R - G \leq 0$ so Right wins going second regardless of Left's move; thus, he wins if Left moves to $A^R - H$.

The moves not covered for Right are those where he moves in G' to one of the new options (such as W) that came from A^{RL}, that is, to some $G - W$. However, since $G - A^R \geq 0$ then Left can win playing second; specifically, she can win if Right moves to $G - W$.

4.37: (page 94) Suppose that $G = H$. Then from Theorem 4.36 we know that the canonical forms of G and H are the same. However, the reductions that we use to obtain canonical form delete parts of the game tree (removing dominated options) or replace parts of it with simpler trees (reversing). In particular, the birthday of any game is at least as great as the birthday of its canonical form, and so it's impossible that there could be a game equal to G whose birthday was less than the birthday of the canonical form of G.

4.39: (page 94) We will show the first assertion, that for any G^L there exists an $H^L \geq G^L$. A symmetric argument shows the second assertion.

Since $G = H$, Left wins moving second on $H - G$. Left's winning response to $H - G^L$ cannot be to $H - G^{LR}$, for then $H - G^{LR} \geq 0$, and since $G = H$ we have $G^{LR} \leq G$ and G had a reversible option. Hence, the winning move must be to some $H^L - G^L \geq 0$, and we have $H^L \geq G^L$.

4.40: (page 94) Since G is a second-player win, $o_R(G^L) = \odot$. Therefore, there is some $G^{LR} \leq 0$, and so G^L is reversible.

4.49: (page 96) Let $G = \{\{4 \mid \{1, \{2 \mid 0\} \mid 0\}\} \mid 3\}$. Now $G - \{4 \mid \{1, \{2 \mid 0\} \mid 0\}\}$ is a left win: (i) Left playing first moves to $G - 0$ and Right only has the move to 3; (ii) Right playing first can move to $3 - \{4 \mid \{1, \{2 \mid 0\} \mid 0\}\}$ and Left plays to $3 - 0$; or (iii) Right playing first can move to $G - \{1, \{2 \mid 0\} \mid 0\}$ and Left plays to $G - 0$ to win. This shows that $\{4 \mid \{1, \{2 \mid 0\} \mid 0\}\}$ is a reversible option, where $\{1, \{2 \mid 0\} \mid 0\}$ is the reversing option and the replacement set is $\{1, \{2 \mid 0\}\}$. Therefore, $G = \{3, \{1, \{2 \mid 0\} \mid 0\} \mid 3\}$. Since $3 > \{1, \{2 \mid 0\} \mid 0\}$, we have that $G = \{3 \mid 3\}$.

Solutions for Chapter 5

Prep 5.1: (page 104)

$$0 \xrightarrow{4T} 4 \xrightarrow[\;3T\;]{\;R\;} {\textstyle -\frac{1}{4} \atop \frac{2}{3}} \xrightarrow[\;R\;]{\;2T\;} {\textstyle \frac{7}{4} \atop -\frac{3}{2}} \xrightarrow[\;2T\;]{\;R\;} {\textstyle -\frac{4}{7} \atop \frac{1}{2}} \xrightarrow[\;R\;]{\;T\;} {\textstyle \frac{3}{7} \atop -2} \xrightarrow[\;2T\;]{\;R\;} {\textstyle -\frac{7}{3} \atop 0}$$

5.1: (page 106) Since $n = \{n-1 \mid \}$, we have that $-n = \{\mid -(n-1)\}$.

5.2: (page 106)

$$\boxminus = \{0 \mid \} = 1;$$

and

$$\boxminus = \{\boxminus \mid \} = \{1 \mid \} = 2.$$

5.4: (page 107) We will do the two examples in two different ways:

$$
\begin{aligned}
\text{"}n+1\text{"} &= \{\text{"}n-1\text{"} \mid \} + \{0 \mid \} && \text{(definition of games } n \text{ and 1)}\\
&= \{\text{"}n-1\text{"} + \text{"}1\text{"}, \text{"}n\text{"} + \text{"}0\text{"} \mid \} && \text{(definition of } +)\\
&= \{\text{"}n\text{"}, \text{"}n\text{"} + 0 \mid \} && \text{(by induction}\\
& && \text{and definition of game 0)}\\
&= \{\text{"}n\text{"} \mid \}.
\end{aligned}
$$

For the second example, we can show that "n" $-$ "1" $=$ "$n-1$" by confirming that the second player wins the difference game "$n-1$" $-$ "n" $+$ "1." All moves have an easy response to 0; for one case (Left's first move on "$n-1$"), you need to observe that "$n-2$" $-$ "$n-1$" $+$ "1" is 0 by induction.

5.7: (page 107) By Theorem 5.5, $A+(-B)+0 \geq 0$ *if and only if* $a+(-b)+0 \geq 0$.

5.8: (page 107) There is no incentive moving from 0. From $n = \{n-1 \mid \}$ Left's incentive is $(n-1) - n = -1$ and Right has no move and so no incentive.

5.10: (page 108) Each component is of one color, and the value is $3 - 3 - 2 + 3 = 1$.

5.11: (page 108) $\frac{1}{4} = \{0 \mid \frac{1}{2}\}$ and $\frac{1}{8} = \{0 \mid \frac{1}{4}\}$.

5.14: (page 109)

$$\frac{15}{16} + \frac{1}{4} = \left\{\frac{7}{8} \mid 1\right\} + \left\{0 \mid \frac{1}{2}\right\}$$

$$= \left\{\frac{7}{8} + \frac{1}{4}, \frac{15}{16} + 0 \mid 1 + \frac{1}{4}, \frac{15}{16} + \frac{1}{2}\right\}$$

$$= \left\{\frac{9}{8}, \frac{15}{16} \mid \frac{5}{4}, \frac{23}{16}\right\}$$

$$= \left\{\frac{9}{8} \mid \frac{5}{4}\right\}$$

$$= \left\{\frac{9}{8} \mid \frac{10}{8}\right\}$$

$$= \frac{19}{16}.$$

Or, confirm that the second player wins on $\frac{15}{16} + \frac{1}{4} - 1\frac{3}{16}$. Moves on $\frac{15}{16}$ are matched by moves on $1\frac{3}{16}$, leaving 0. A move on $\frac{1}{4}$ is worse, and a response in either of the other components leaves the second player in a favorable position.

5.15: (page 109) Since $\frac{m}{2^j} = \{\frac{m-1}{2^j} \mid \frac{m+1}{2^j}\}$, the left incentive is $\frac{m-1}{2^j} - \frac{m}{2^j} = -\frac{1}{2^j}$ and the right incentive is $\frac{m}{2^j} - \frac{m+1}{2^j} = -\frac{1}{2^j}$.

5.24: (page 111) Note that $\pm n = \{n \mid -n\}$ is in canonical form and both n and $-n$ have birthdays n. Therefore, $\pm n$ has birthday $n+1$ by Definition 4.1. Then,

$$\pm\left(n + \frac{i}{2^j}\right) = \left\{n + \frac{i}{2^j} \mid -n - \frac{i}{2^j}\right\}$$

and

$$n + \frac{i}{2^j} = \left\{n + \frac{2i-1}{2^{j-1}} \mid n + \frac{2i+1}{2^{j-1}}\right\}.$$

So, n has birthday n and, by induction, $n + \frac{i}{2^j}$ has birthday $n+j$. Therefore, $\pm\left(n + \frac{i}{2^j}\right)$ has birthday $n+j+1$.

5.26: (page 112)

(a) $\{\frac{1}{2} \mid 2\} = 1$.

(b) $\{\frac{1}{8} \mid \frac{5}{8}\} = \frac{1}{2}$.

(c) $\{-1\frac{27}{64} \mid -1\frac{9}{32}\} = -1\frac{3}{8}$.

5.31: (page 116) The incentives are all equal to $-\frac{1}{4}$, and either player would move the leftmost piece possible in any summand to achieve this incentive.

5.33: (page 117)

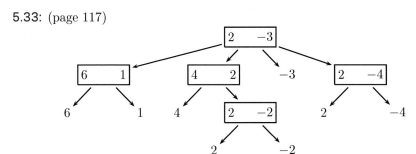

5.36: (page 118) Since $\mathbf{RS}(G) = 1 > -2 = \mathbf{LS}(H)$, we have that $G > H$.

5.37: (page 118) Blindly applying the definition of stops gives $\mathbf{LS}(G) = 1$ and $\mathbf{RS}(G) = 2$, contrary to Theorem 5.34, so the only conclusion is that, since all options are in canonical form, G is not. In fact, it is not hard to show that $G = 1$.

5.47: (page 120) The first player wins on $\uparrow + *$: Left moves on the second component leaving $\uparrow + 0$, while Right moves on the first component leaving $* + * = 0$.

$\uparrow + \uparrow + * > \uparrow + *$, so Left still wins moving first. Right, however, has two moves both of which lose, one to $\uparrow + \uparrow + 0 > 0$ and one to $* + \uparrow + * = \uparrow > 0$.

5.49: (page 124) Let $\blacksquare\blacksquare^{n}$ denote the CLOBBER position $\boxed{\blacksquare}\boxed{\blacksquare}\boxed{\blacksquare}\boxed{\blacksquare}\boxed{\blacksquare}\cdots\boxed{\blacksquare}$ with n blue pieces. Then,

$$\blacksquare\blacksquare^{2n} = \left\{0 \mid \blacksquare\blacksquare^{2n-1}\right\} = \left\{0 \mid (2n-2)\!\cdot\!\uparrow *\right\} = (2n-1)\!\cdot\!\uparrow *$$

and

$$\blacksquare\blacksquare^{2n+1} = \left\{0 \mid \blacksquare\blacksquare^{2n}\right\} = \left\{0 \mid (2n-1)\!\cdot\!\uparrow *\right\} = 2n\!\cdot\!\uparrow *.$$

In other words,

$$\blacksquare\blacksquare^{n} = n \cdot \uparrow* + *,$$

where the latter should be parsed as $n \cdot (\uparrow*) + *$.

5.50: (page 124) On $a + \{x \mid x\} - \{y \mid z\}$, if the first player moves on one switch, the opponent moves on the other switch leaving either $a - x - z = 0$ or $a + x - y = 0$. If the first player moves on a, the second player plays on switches until both are resolved, leaving a favorable number.

5.51: (page 124) $\pm x \pm x = 0$ since it is a second-player win.

5.52: (page 124)

(b) Left moving first gets a maximum total of $a + x - y + z$, so she wins if $a > -x + y - z$. Left moving second gets a maximum total of $a - x + y - z$ and so wins if $a \geq x - y + z$. @par

(c) Left, moving first, wins $G = a \pm x_1 \pm x_2 \pm x_3 \cdots \pm x_n$ if $a + x_1 - x_2 + x_3 \cdots \pm x_n > 0$ and wins moving second if $a - x_1 + x_2 - x_3 \cdots \pm x_n \geq 0$.

5.53: (page 124) Denote the position by (m, n, p).

Claim: $(m, n, p) = \{m - 1 \mid \{p - 1 \mid -(n - 2)\}\}$.

If Left moves to eliminate the red dominoes, then the best option is $m - 1$. Her other options are to positions of values $\{m' - 1 \mid \{p - 1 \mid -(n - 1)\}\}$ or $\{m - 1 \mid \{p' - 1 \mid -(n - 2)\}\}$. However, Left, playing second, wins $(m - 1) - \{m' - 1 \mid \{p - 1 \mid -(n - 2)\}\}$, so this option is dominated. Left's option $\{m - 1 \mid \{p' - 1 \mid -(n - 1)\}\}$ reverses to $p' - 1$, which is also dominated. Right's options are to $\{m - 1 \mid -(n' - 1\}$ and $\{p - 1 \mid -(n' - 1)\}$ for $n' \leq n$. All these are dominated by $\{p - 1 \mid -(n - 1)\}$. Therefore,

$$(m, n, p) = \{m - 1 \mid \{p - 1 \mid -(n - 2)\}\}.$$

5.55: (page 126)   Each of the $\lfloor \frac{s}{2} \rfloor$ moves by Left gains $n_\bullet$ moves for the remaining elephants. (Similarly for Right.) In total, that changes the difference in available non-central moves by $\lfloor \frac{s}{2} \rfloor (n_\bullet - n_\circ)$.

5.57: (page 127) The left and right incentives of moving a central beast on a component with $n + 2$ beasts is $\{n \mid 0\}$. The incentive of moving a non-central beast is -1. The incentive $\{n \mid 0\}$ with maximal n dominates.

5.58: (page 127) Let x be a positive number and consider $x - {+}_G = x + {\leftarrow}_G$. Left wins by playing to $x + \{G \mid 0\}$, for Right's response to $x > 0$ fails immediately, and his response to $x^R + \{G \mid 0\}$ fails when Left counters to $x^R + G$, which is also positive since $x^R > x$ is also a positive number.

Right moving first in $x + {}_G$ can move to $x > 0$ or to $x^R + {}_G$, which is also positive by induction since $x^R > x$ and is simpler than x.

5.60: (page 128) Consider $_y - k \cdot {}_x = {}_y + k \cdot {}_x$. Left playing always moves in one of the $_x$ to $\{x \mid 0\}$. If Right fails to respond, then Left's move to x dominates all other positions. (As in the proof of Theorem 5.59, Left plays arbitrarily on the other positions, ignoring x until it is the only game remaining. If, in the meanwhile, Right has played on x, that play only increases x.) After all of the $_x$ summands have been played out, it is Left's move in either $_y$ or $\{0 \mid -y\}$, both of which she wins.

5.66: (page 131) The position is $c+2$ red dominoes sandwiched between $a+1$ blue dominoes at one end and $b+1$ blue dominoes at the other end. For example,

$$\text{[dominoes]} = \left\{5 \,\middle\|\, 3 \mid -1\right\}.$$

5.67: (page 131)

$$\frac{1}{2} = \{0 \mid 1\} = \text{[dominoes]}$$

$$\frac{1}{4} = \left\{0 \mid \tfrac{1}{2}\right\} = \text{[dominoes]}$$

$$\frac{3}{8} = \left\{\tfrac{1}{4} \mid \tfrac{1}{2}\right\} = \text{[dominoes]}$$

5.71: (page 133) This is covered by the induction. In the second paragraph, we only needed to show how Left wins in response to Right's legal moves. (When $a = b = c = 0$, there are no legal moves, so no worries.)

5.73: (page 133) From Lemma 5.16, $A + B - C = 0$ if and only if $a + b - c = 0$. Similarly, $A - B + 0 \geq 0$ if and only if $a - b + 0 \geq 0$.

Solutions for Chapter 6

Prep 6.1: (page 138)

(a)

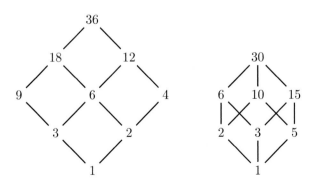

(b) The gcd of a and b is found by looking at the highest element that has paths up to both a and b. (Note that the two paths will only meet at the gcd, for any higher meeting point would be a common divisor that is even greater.) The lcm of a and b is found by looking at the lowest element that has paths down to both a and b.

Prep 6.2: (page 138)

(a)

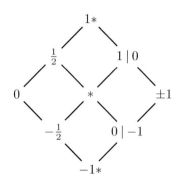

(b) If one (or both) sets of option can be empty, we would also include the games 2, 1, −1, and −2.

6.2: (page 141) Since $\{0, * \mid *\} > 0$, the left option to $*$ reverses out through 0. Since $\{0, * \mid 1\} > 0$ and $\{0, * \mid \} > 0$, the same is true in the other two examples.

6.8: (page 144) The statement that we are trying to prove is of the form, "For any infinitesimal G ....." Inductively, we can only assume that the statement is true for smaller *infinitesimal* games.

6.10: (page 145) To the contrary, suppose that $\mathbf{LS}(G^{LR}) > \mathbf{LS}(G)$ for all G^{LR}. By the definition of stops, $\mathbf{RS}(G^L)$ is defined as the maximum $\mathbf{LS}(G^{LR})$, and so $\mathbf{RS}(G^L) > \mathbf{LS}(G)$. But $\mathbf{LS}(G)$ is the maximum among $\mathbf{RS}(G^{L'})$ of which $\mathbf{RS}(G^L)$ is one. So, $\mathbf{LS}(G) \geq \mathbf{RS}(G^L) > \mathbf{LS}(G)$, which is a contradiction.

6.11: (page 145) $\mathbf{RS}(G^R) \leq \mathbf{LS}(G^R)$ and $\mathbf{RS}(G) \leq \mathbf{LS}(G)$ by Theorem 5.34. $\mathbf{LS}(G^R) = \mathbf{RS}(G)$ since G^R was chosen to be the right option from G with minimal left stop, which by definition is $\mathbf{RS}(G)$. Lastly, $\mathbf{LS}(G) \leq 0$ was assumed in the statement of the theorem.

6.16: (page 147) Suppose that $\mathbf{LS}(G) = x = \mathbf{RS}(G)$. By repeated application of Theorem 6.13, $\mathbf{LS}(G - x) = 0 = \mathbf{RS}(G - x)$, and so $G = x + (G - x)$ is a number plus an infinitesimal.

6.21: (page 148) We use the definitions given in the proof; the partial order diagram confirms the results:

(a) $* \vee 0 = \{0 \mid \lceil * \rceil \cap \lceil 0 \rceil\} = \{0 \mid \{0,1\} \cap \{*,1\}\} = \frac{1}{2}$;

(b) $* \vee \frac{1}{2} = \{0,0 \mid \lceil * \rceil \cap \lceil \frac{1}{2} \rceil\} = \{0 \mid \{0,1\} \cap \{1\}\} = \frac{1}{2}$;

(c) $\uparrow \vee \pm 1 = \{0,1 \mid \lceil \uparrow \rceil \cap \lceil \pm 1 \rceil\} = \{1 \mid \{*,1\} \cap \{-1,0,*,1\}\} = \{1 \mid *\}$;

(d) $\uparrow \wedge \pm 1 = \{\lfloor \uparrow \rfloor \cap \lfloor \pm 1 \rfloor \mid *, -1\} = \{\{0,*,-1\} \cap \{1,0,*,-1\} \mid -1\} = \{0,* \mid -1\}$.

6.23: (page 150) For the diagram on the left,

$$a \wedge (b \vee c) = a \wedge 1 = a;$$
$$(a \wedge b) \vee (a \wedge c) = 0 \vee 0 = 0.$$

For the right,

$$a \wedge (b \vee c) = a \wedge 1 = a;$$
$$(a \wedge b) \vee (a \wedge c) = b \vee 0 = b.$$

6.25: (page 151) On $\uparrow - n \cdot \alpha$, Left moving first attacks the αs one by one. Right must either reply locally or on another α; after two moves, one or two αs have been converted to $* = \{1* \mid \{0 \mid -1\}\}$. (If Right moves on $\uparrow$, Left grabs 1*, an overwhelming advantage.) After all αs are gone, Left wins moving first on $\uparrow$ or $\uparrow*$.

Solutions for Chapter 7

Prep 7.1: (page 154)

Decimal	Binary
$-2\frac{1}{2}$	-10.1
23	10111
21	10101
$\frac{1}{4}$	.01
$\frac{5}{8}$	.101
$\frac{13}{32}$	.01101
$-12\frac{9}{16}$	-1100.1001
$-3\frac{21}{32}$	-11.10101

7.1: (page 155) The second player can play the Tweedledum-Tweedledee strategy and apply induction. That is, whatever option the first player chooses in one component, the second player makes the same move in the other component.

7.2: (page 155) If G and H are impartial games and $G \leq H$, then $G = H$ since $H - G$ is impartial and cannot be positive. Therefore, the dominated or reversible left options of an impartial game are likewise dominated or reversible as right options. So, by induction, the canonical form has the same left and right options.

7.5: (page 157) $*i \parallel *j$ because one is both a left and right option from the other. (That is, if $i > j$, the first player moves $*i - *j$ to $*j - *j = 0$ and wins.) This directly implies that there are no dominated nor reversible options.

7.9: (page 157)

(a) We find $\text{mex}\{0, 3, 4, 8, 2\} = 1 = \text{mex}\{0, 6, 4\}$, so the game is $*$. @par

(b) From $G - *$, if the first player moves to $G - 0$ or $0 - *$, the second can reply to $0 - 0$. If, on the other hand, the first player moves to $*j - *$ for $j > 1$, the second player replies to $* - * = 0$.

7.11: (page 158) By definition, the nim-sum addition in the kth binary place (or column) is independent from the addition in other binary places. Furthermore, since addition is commutative and associative, so is the columnwise nim-sum, for it is just the parity of the sum.

For $a \oplus a$, in each binary place we have $0 + 0 = 0$ or $1 + 1 = 0$.
If $a \oplus b = c$ then

$$a \oplus b \oplus c = (a \oplus b) \oplus c = c \oplus c = 0.$$

If $0 = a \oplus b \oplus c$ then

$$c = 0 \oplus c = (a \oplus b \oplus c) \oplus c = a \oplus b \oplus (c \oplus c) = a \oplus b \oplus 0 = a \oplus b.$$

7.14: (page 160)

(a) $3 \oplus 5 \oplus 7 = 1$; all heaps are odd, so remove 1 from any heap.

(b) $2 \oplus 3 \oplus 5 \oplus 7 = 3$, whose leftmost bit is second from the right. In binary, 2, 3, and 7 have a leftmost bit there, so there is a winning move on each of these heaps. Move $2 \to 1$ or $3 \to 0$ or $7 \to 4$.

(c) $2 \oplus 4 \oplus 8 \oplus 32 \oplus = 46$; the only winning move is to reduce the heap of 32 to $32 \oplus 46 = 14$.

(d) $2 \oplus 4 \oplus 10 \oplus 12 = 0$, so there is no winning move.

7.17: (page 160) Suppose that a player moves to 0 (according to NIM strategy). Whenever the opponent increases the size of a heap (drawing down the bag), one of the moves dictated by the NIM strategy is simply to pocket those same counters. So, there exists a move back to 0, winning by induction. (Eventually the bag runs dry, and the game reverts to NIM.)

7.18: (page 161)

(a) As in Exercise 7.17, the player who sees a position with nim-sum 0 can win by always removing counters; eventually, his opponent will only be able to make reducing moves.

(b) The game is technically loopy since both players on their first move could take one counter and then on their second moves add one counter and so on. However, the first part of the exercise shows that the player who is in a winning position never has to add counters to win.

7.20: (page 161) The strategy is exactly the same except if the number of coins is odd, the leftmost coin on square i is equivalent to a NIM heap of size i.

7.23: (page 162) The nim-value of G is the mex of the nim-values of its options. Since there are only n options, at least one of the values $\{0, 1, 2, \ldots, n\}$ does not occur as the nim-value of an option. Hence, $\mathcal{G}(G) \leq n$.

7.27: (page 163)

$$\mathcal{G}(n) = \begin{cases} 0 & \text{if } n \equiv 0 \pmod 5, \\ 1 & \text{if } n \equiv 1 \pmod 5, \\ 0 & \text{if } n \equiv 2 \pmod 5, \\ 1 & \text{if } n \equiv 3 \pmod 5, \\ 2 & \text{if } n \equiv 4 \pmod 5. \end{cases}$$

7.30: (page 165)

1231451671...	purely sapp regular, $p = 3$, $s = 2$, $S = \{1, 2\}$;
1123123123...	periodic, $p = 3$, pre-period 1;
1122334455...	purely arithmetic periodic, $p = 2$, $s = 1$;
0123252729...	sapp regular, $p = 2$, $s = 2$, $S = \{1\}$, pre-period 01;
0120120120...	purely periodic, $p = 3$;
1112233445...	arithmetic periodic, $p = 2$, $s = 1$, pre-period 1.

7.32: (page 167) The sequence is $0001112220331\dot{4}\dot{0}0201313\dot{2}$.

7.35: (page 168) Once you have gone as far as 0011220310210210210 you can be confident the sequence is $0011220310\dot{2}$. We have, $l = 8$, $p = 3$, and $a = 7$, so it suffices to inspect $\mathcal{G}(n)$ for $n \in \{8, \ldots, 17\}$. In particular, confirm $\mathcal{G}(n) = \mathcal{G}(n + 3)$ for the 7 values $8 \leq n \leq 14$.

7.37: (page 169) The first 20 values ($n = 0$ through $n = 19$) are

$$00110213223344554657.$$

7.42: (page 171) Suppose that the sequence (7.1) matches term-by-term for the two values $n = n_0$ and $n = n_0'$ for $n_0' > n_0$. Define $p = n_0' - n_0$ and fix $n_0' \le n \le n_0' + 2a$. Then, adding equations from (7.1) form two telescoping sums, and we have

$$G(n_0 + 1) - \quad G(n_0) \quad = G(n_0' + 1) - \quad G(n_0');$$
$$G(n_0 + 2) - G(n_0 + 1) = G(n_0' + 2) - \quad G(n_0' + 1);$$
$$\vdots$$
$$G(n) \quad - G(n - 1) = G(n + p) - G(n + p - 1).$$

Adding these equations yields $G(n + p) - G(n_0') = G(n) - G(n_0)$, and so $G(n + p) - G(n) = G(n_0') - G(n_0) = s$.

Or, use induction:

$$\mathcal{G}(n + p) = \mathcal{G}(n + p - 1) + (\mathcal{G}(n + p) - \mathcal{G}(n + p - 1))$$
$$= s + \mathcal{G}(n - 1) + (\mathcal{G}(n) - \mathcal{G}(n - 1)) \quad \text{(by induction and (7.1))}$$
$$= s + \mathcal{G}(n).$$

In the base case, when $n = n_0$, $\mathcal{G}(n_0 + p) = \mathcal{G}(n_0') = s + \mathcal{G}(n_0)$ by definition of s.

7.44: (page 172) The sequence has period 12 and saltus 4: $\dot{0}0110213223\dot{3}(+4)$. Since $a = 4$, $p = 12$, and $l = 0$, it suffices to compute values through $l + 2a + p = 20$ and confirm that $\mathcal{G}(n) = \mathcal{G}(n - p)$ for $12 \le n \le 20$.

7.45: (page 172) It suffices to compute through $n = l + 2a + p = 8 + 2 \cdot 10 + 3 = 31$.

7.47: (page 175) The next values are 4, 2, 6, 4, 1.

7.51: (page 177) Corollary 7.50 says that we need to run out as far as pile sizes $l + s - 1$, i.e., $27 + 9 - 1 = 35$, $129 + 1 - 1 = 129$, and $117 + 15 - 1 = 131$, respectively.

Solutions for Chapter 8

Prep 8.1: (page 180)

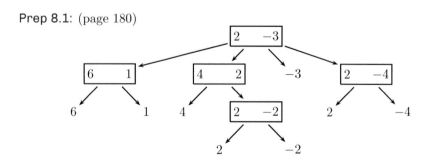

8.4: (page 183) The bottom game has **LS** = 2 and **RS** = −2. The middle left game has **LS** = 4 and **RS** = 2. The middle right game has **LS** = −3 and **RS** = −5. The top game has **LS** = 3 and **RS** = −3.

8.8: (page 188) The values for a single heap are periodic, with period 4; specifically, heap size $n \equiv 0 \pmod 4$ has value 0, heap size $n \equiv 1 \pmod 4$ has value 1, heap size $n \equiv 2 \pmod 4$ has value $\{1 \mid 0\}$, and heap size $n \equiv 3 \pmod 4$ has value $\{1, \{1 \mid 0\} \mid 0\}$. The mean value of $\{1 \mid 0\}$ is $\frac{1}{2}$ as is the mean value of $\{1, \{1 \mid 0\} \mid 0\}$. Therefore, the disjunctive sum of 6 heaps of size 10, 20 of size 11, and 132 of size 31 has approximate value $6(\frac{1}{2}) + 20(\frac{1}{2}) + 132(\frac{1}{2}) = 78$.

Note: For this game, $\{1 \mid 0\} + \{1 \mid 0\} = 1$ and $4 \cdot \{1, \{1 \mid 0\} \mid 0\} = 2 + X$, where X is a *tiny*, i.e., $X = \{0 \mid \{0 \mid Y\}\}$.

8.10: (page 194) If none of X_t, Y_t, and $(X + Y)_t$ is a number, then

$$X_t + Y_t = \{X_t + Y^L{}_t - t, X^L{}_t + Y_t - t \mid X_t + Y^R{}_t + t, X^R{}_t + Y_t + t\}$$
$$= \{(X + Y^L)_t - t, X^L + Y_t - t \mid (X + Y^R)_t - t, (X^R + Y_t - t\}$$
$$= (X + Y)_t.$$

8.15: (page 197)

(a) P has temperature 1 and Q has temperature 2. So, both players are advised to move on Q first. However, when Left moves first, she can do better by moving on P, for Right must immediately respond locally. @par

(b) In this example, Thermostrat gives the same advice as Hotstrat. In particular, the compound thermograph of P and Q is maximally left at temperature 0, and at that temperature Q's thermograph is widest, the difference between left and right stops being 4.

8.16: (page 197) The goal of a strategy is to do well against *any* opponent, and not just an opponent following the same strategy. If your opponent makes a low-temperature move, that should not oblige you to make poor moves for the remainder of the game.

8.19: (page 198) $\frac{1}{2} \cdot 1* = \{1 \mid *\}$,
$\frac{1}{4} \cdot 1* = \{1 \mid \{0 \mid -1*\}\}$,
$\frac{1}{8} \cdot 1* = \{1 \mid \{0 \mid \{-1 \mid -2*\}\}\}, \dots .$

8.21: (page 200) In each step of the induction, the birthdays of both game sums $A + B + C$ and $A \cdot U + B \cdot U + C \cdot U$ (perhaps $- t$) always decrease.

8.23: (page 201)

$$* \int \frac{11}{4} + * \int \frac{5}{4} - * \int \frac{9}{2} = * \int -1/2 = \{0 \mid -1/2*\}$$

Solutions for Chapter 9

Prep 9.1: (page 206) See Section 5.4.

9.4: (page 209) Since $*m = -*m$, $\uparrow *n > *m$ *if and only if* $\uparrow *(n \oplus m) > 0$. By Lemma 9.2, this is equivalent to $n \oplus m \neq 1$ or equivalently $n \oplus 1 \neq m$.

9.7: (page 210) By induction, the left options $(a - j, a + i + j)$ and the right options $(a + i + j, a - j)$ have value $*(a - j)$ for $0 < j \leq a$; therefore,

$$(a, a + i) = \{0, *1, *2, \ldots, *(a - 1) \mid 0, *1, *2, \ldots, *(a - 1)\} = *a.$$

9.8: (page 210) The game is positive, for Left moving first moves to $(1, a+1, 1)$ and moves second to a single heap:

$$
\begin{aligned}
(a, a + i, 1) &= \{(2a + i, 1), (1, 2a + i - 1, 1), \ldots, \\
&\quad (a - 1, a + i + 1, 1)(a, a + i + 1)\} \\
&= \{*, 0, \uparrow *(2 \oplus 1), \ldots, \uparrow *(a - 1 \oplus 1) \mid *a\} \\
&= \{0 \mid *a\} \text{ (since the other left options all reverse out through 0)} \\
&= \uparrow *(a \oplus 1).
\end{aligned}
$$

9.9: (page 210) As in the last exercise, the game is positive and so most of Left's options reverse out:

$$
\begin{aligned}
(a + i, a, 1) &= \{(a - 1, a + i + 1, 1), \ldots, (2, a + i + a - 2, 1), 0, * \mid *(a + 1)\} \\
&= \{\uparrow *(a - 1 \oplus 1), \ldots, \uparrow *(2 \oplus 1), 0, * \mid *(a + 1)\} \\
&= \{0 \mid *(a + 1)\} \\
&= \uparrow *((a + 1) \oplus 1)).
\end{aligned}
$$

9.11: (page 210) In base $\uparrow = \{0 \mid *\}$, $\uparrow^1 = \uparrow$ and $\uparrow^2 = \{0 \mid * - \uparrow\} = \{0 \mid \downarrow *\}$.

9.14: (page 212) Let $H = .i_1 i_2 \ldots$ and suppose without loss of generality that the first non-zero digit i_n is positive. Let m be the sum of the absolute values of all the digits in positions to the right of i_n. Then, $H \geq G^n - m \cdot G^{m+1} > 0$.

9.18: (page 213) From Theorem 9.15,

$$\underbrace{.11\ldots1}_{n+1}* = \left\{\; \underbrace{.11\ldots1}_{n+1},\; \underbrace{.11\ldots1}_{n}* \;\middle|\; *+*,\; \underbrace{.11\ldots1}_{n+1} \;\right\}.$$

Right's move to $*+* = 0$ dominates, while Left's move to

$$\underbrace{.11\ldots1}_{n+1}$$

reverses through $*$ to 0.

9.24: (page 215)

Reflexive: $G - G = 0$ is infinitesimal.

Symmetric: $G - H$ is infinitesimal if and only if $H - G$ is.

Transitive: If $G-H$ and $H-J$ are infinitesimal, so is $(G-H)+(H-J) = G-J$.
(The sum of two infinitesimals is infinitesimal because if $-x < G_1 < x$
and $-x < G_2 < x$, then $-2x < G_1 + G_2 < 2x$.)

9.35: (page 221) To see that $\mathrm{AW}(\boldsymbol{+}_2) = 0$, confirm that $\quad_2 < \uparrow\!\!\text{☆}$. Also,
$\mathrm{AW}(\{0 \mid \quad_2\}) = 1$ for $\quad < \{0 \mid \quad_2\} < \Uparrow$.

9.36: (page 221) Since $.p = p \cdot \uparrow$, $\mathrm{AW}(.p) = p$. $.01$ is infinitesimal with respect
to $.1$ and so must have atomic weight 0. So,

$$\mathrm{AW}(.pq) = \mathrm{AW}(.p + .0q) = \mathrm{AW}(.p) + q\,\mathrm{AW}(.01) = p.$$

Alternatively, since $.01 = \{0 \mid \downarrow\!*\}$, $\mathrm{AW}(.01) = \{-2 \mid 1\} = 0$. (Note that
since $.01 > 0$, with or without invoking the exception, the atomic weight cal-
culus of Theorem 9.39 gives 0.)

9.40: (page 222) Note that $\{n - 1 \mid n + 2\}$ is a number, and we are in the
exceptional case. Two integers fit, those being n and $n + 1$. When $n \geq 1$,
$g > \quad$ and so $\mathrm{AW}(g) = n + 1$. If, however, $n = 0$ or $n = -1$, $g \parallel \quad$ and so
$\mathrm{AW}(g) = 0$. When $n < -1$, $\mathrm{AW}(g) = n$.

9.41: (page 222) The largest integer in $\{0 - 2 \mid 0 + 2\}$ is 1, so it suffices to
confirm that Left wins moving second on $g + \quad$. Left can reply to most Right
moves leaving a game of the form $*p + *p$. However, if Right moves to $g + *p$
for $p > m$, Left replies to $g + *(n+1)$. From there, Right has no choice but to
let Left leave a pair of matching $*$s.

9.42: (page 223)   Either player's move causes ⬚ to shift one square left, and the game ends when it falls off the board. Hence, the game is a variant of SHE LOVES ME SHE LOVES ME NOT. For those fond of formality, by induction,

$$\square^n\,\boxed{\;}\,\boxed{\;} = \left\{\square^{n-1}\,\boxed{\;}\,\boxed{\;} \;\middle|\; \square^{n-1}\,\boxed{\;}\,\square\,\boxed{\;}\right\}$$

$$= \left\{\square^{n-1}\,\boxed{\;}\,\boxed{\;} \;\middle|\; \square^{n-1}\,\boxed{\;}\,\boxed{\;}\right\}$$

$$= \{0 \mid 0\} \text{ or } \{* \mid *\}$$

$$= * \text{ or } 0.$$

9.43: (page 223)

$$\square^m\,\boxed{\;}\,\square^n\,\boxed{\;}\,\boxed{\;}$$

$$= \left\{\square^{m-1}\,\boxed{\;}\,\square^{n+1}\,\boxed{\;}\,\boxed{\;}\,,\right.$$

$$\left.\square^{m-1}\,\boxed{\;}\,\square^n\,\boxed{\;}\,\boxed{\;}\,\square^{m-1}\,\boxed{\;}\,\square^n\,\boxed{\;}\,\square\,\boxed{\;}\right\}$$

$$= \{m\cdot\uparrow + (n+m)\cdot *, \; m\cdot\uparrow + (n+m-1)\cdot * \mid m\cdot\uparrow + (n+m-1)\cdot *\}$$

$$= \{m\cdot\uparrow + *, \; m\cdot\uparrow \mid m\cdot\uparrow + (n+m-1)\cdot *\}$$

$$= \{0 \mid m\cdot\uparrow + (n+m-1)\cdot *\} \qquad \text{(reversible options)}$$

$$= (m+1)\cdot\uparrow + (n+m)\cdot *.$$

9.44: (page 224)

$$\square^m\,\boxed{\;}\,\square^n\,\boxed{\;}\,\boxed{\;}$$

$$= \left\{\square^{m-1}\,\boxed{\;}\,\square^n\,\boxed{\;}\,\boxed{\;}\,\square^{m-1}\,\boxed{\;}\,\square^{n+1}\,\boxed{\;}\,\boxed{\;}\,,\right.$$

$$\left.\square^{m-1}\,\boxed{\;}\,\square^n\,\boxed{\;}\,\square\,\boxed{\;}\right\}$$

$$= \{m\cdot\downarrow + (n+m-1)\cdot * \mid m\cdot\downarrow + (n+m)\cdot *, \; m\cdot\downarrow + (n+m-1)\cdot *\}$$

$$= (m+1)\cdot\downarrow + (n+m)\cdot * \qquad \text{(as in the last exercise).}$$

9.45: (page 224) If either end is green, or if the ends have opposite colors, the first player wins and the game is incomparable with zero. If the ends are both blue (respectively, red) the game is positive (respectively, negative).

9.49: (page 226) The game lasts two moves and the second player wins.

Solutions for Chapter 10

10.3: (page 235) Certainly $G \sim_\epsilon G$ since $G - G = 0$, which is infinitesimal. If $G \sim_\epsilon H$ and $H \sim_\epsilon K$, then $G - H$ and $H - K$ are both infinitesimals. Since the sum of two infinitesimals is also infinitesimal $G - K = (G - H) + (H - K)$ is infinitesimal, and so $G \sim_\epsilon K$. Finally, if $G \sim_\epsilon H$ then $G - H$ is infinitesimal. But, the negative of an infinitesimal is infinitesimal, so $H - G$ is infinitesimal and $H \sim_\epsilon G$. The arguments for $\geq_\epsilon$ are very similar.

10.17: (page 237) A move in G for, say, Left involves sliding the coin one or more spaces down and to the left. Any single move could be followed by another move of the same type if there are spaces remaining — but the result would be the same as if just one move (whose length was the sum of the two lengths chosen) had been made. Since the same applies for Right, and since any option of a MAZE position is also a MAZE position, the condition $H^{\mathcal{L}\mathcal{L}} \subseteq H^{\mathcal{L}}$ and $H^{\mathcal{R}\mathcal{R}} \subseteq H^{\mathcal{R}}$ for all followers H of G is satisfied. Thus, MAZE is hereditary-transitive.

10.22: (page 240) If we play NIM in a single heap of size $n + m$, we could view the counters as being arranged from left to right in a row and always remove a block of counters from the right-hand end. If we remove m or fewer counters, we get "a row of n and some more," while if we remove more than m counters, we get "a row of fewer than n." These options correspond naturally to the options of $*n : *m$ and vice versa.

Solutions for Appendix A

A.2: (page 260) The ith positive odd integer is $2i - 1$. Implicit in the statement is that $n \geq 0$, for these are the only values of n for which "the first n ..." makes any sense.

$$\sum_{1 \leq i \leq n} (2i - 1) = (2n - 1) + \sum_{1 \leq i \leq n-1} (2i - 1)$$
$$= (2n - 1) + (n - 1)^2 \quad \text{(by induction)}$$
$$= (2n - 1) + (n^2 - 2n + 1) = n^2.$$

For the base case, when $n = 0$,

$$\sum_{1 \leq i \leq 0} (2i - 1) = 0,$$

for any empty sum is 0.

A.4: (page 261) Let $a = n \pmod{11}$.

(a) If $a = 0$, all legal moves leave a pile that is not equivalent to 0, and so lose by induction. Since all moves lose, the second player wins. If, on the other hand, $a \neq 0$, $n - a \equiv 0$, and so removing a coins wins by induction.

(b) No base case is required since the first sentence, "If $a = 0$, all legal moves leave a pile that is not equivalent to 0," handles $n = 0$ properly. "All legal moves ..." is vacuously satisfied.

Solutions for Appendix B

B.1: (page 274)

```
> G := 2||1|*
2||1|*
> G+G
3|2*
> G+G+G+G
5*
> G.Mean
5/4
> G.Temperature
3/4
> G>1
true
> G<>1+*
true
>
```

B.2: (page 274)

```
> ^^^. LeftOptions
0
> ^^^. RightOptions
^^*
```

Thus, the canonical form of ⇑ is $\{0 \mid ⇑*\}$.

B.3: (page 274) H is infinitesimal and is bounded between $2 \cdot {\rightarrowtail}_2 < H < 2 \cdot {+}_2$:

```
> H := game.grid.Domineering(Grid(4,4)).CanonicalForm
+-(0,2|0,2+Tiny(2)|2|0,Miny(2))
```

```
> H < 1/128
true
> H < 2.Tiny
false
> H < 2.Tiny + 2.Tiny
true
```

B.4: (page 274) You can enter something like:

```
> a:=5; b:=3; c:=2; a|b+c.Tiny == a+c.Tiny | b+c.Tiny
false
> a=5; b=2; c=2; a|b+c.Tiny == a+c.Tiny | b+c.Tiny
true
```

In the worksheet, you can edit the line and hit enter again, so you need not type the expression over and over. Once you learn to create functions, you can type:

```
> f := (a,b,c) -> a|b+c.Tiny == a+c.Tiny | b+c.Tiny
<Object of type Procedure>
> f(5,3,2)
false
> f(5,2,2)
true
```

B.5: (page 274)

```
> H := game.grid.Domineering(Grid(4,4))
> H.SensibleLeftLines
```

(Of course, Right's lines are symmetric.)

B.6: (page 274)

```
> G := game.grid.Clobber("xoxo|oxx.|xox.")
> G.SensibleLeftLines
> G.SensibleRightLines
> G.AtomicWeight
```

B.7: (page 275)

```
> C := game.grid.Clobber("xxxxxxx|xoxxxxx|xxxxxxx")
> C.CanonicalForm
> C.AtomicWeight
```

Appendix D
Rulesets

This appendix contains a glossary of the rules to many of the games appearing in the book. If a game only made a brief appearance in a problem or example, then it may not appear here. In most sample games, Left moves first, but the choice was arbitrary as either player can start from any position. Moves in the sample games were chosen to illustrate the rules and do not necessarily represent good play! Unless otherwise noted, the winner is determined by *normal play;* that is, the player who makes the last legal play wins.

We use the following standing conventions throughout, without explicit comment:

- Blue or black pieces, vertices, etc. belong to Left; red or white pieces, vertices, etc. belong to Right.

- In games involving moving pieces on a one-dimensional strip, Left typically moves from left to right while Right moves from right to left.

- If we only specify Left's moves, then Right's moves are similar allowing for appropriate reversals of color and/or direction.

Many games are played on a "board," which amounts to an underlying graph — almost always a path or a grid. Such games often have variants where the underlying graph is more general, but we do not usually mention such variants explicitly.

ALL-SMALL *ruleset*

> The ALL-SMALL variant of a *ruleset* is typically obtained by declaring that play in a component ends when either player has no legal move in *ruleset*.

ALLBUT(S)

> See SUBTRACTION$(L \mid R)$.

AMAZONS

Position: A rectangular board with pieces called *amazons* (either blue or red) and destroyed squares. The standard starting position is a 10×10 board with red amazons in the fourth and seventh squares of the bottom row and on the left and right edges of the fourth row, with blue amazons positioned symmetrically in the upper half of the board.

Moves: A player's amazon moves like a chess queen — any number of squares in a horizontal, vertical, or diagonal line. Once it gets to its final position, the amazon throws an arrow that also moves like a chess queen. The arrow destroys the board square it lands in. Both the amazon and the arrow it throws cannot cross or enter a destroyed square or a square occupied by an amazon of either color.

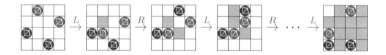

BOXCARS

Position: A finite strip of squares, each square is empty or occupied by a blue or red *boxcar*. Contiguous strings of boxcars must all be of the same color.

Moves: A boxcar occupies one square; a *train* is any maximal string of contiguous pieces (including just a single boxcar) all of the same color. Left chooses a blue train and moves it one square to the right. If this results in a longer string of contiguous boxcars, then every boxcar in the string turns blue. Right moves leftward. Trains cannot be moved off the end of the board.

Variants: In ALL-SMALL BOXCARS the game is over when only pieces of one color remain on the board.

BRUSSELS SPROUTS

Position: A collection of vertices in the plane with four arms emanating from each vertex. Some of these arms may be connected by non-crossing edges. No arm may be connected to itself.

Moves: Draw a curve joining one arm to another that does not cross any other curve or pass through any other vertex. On this curve place another vertex with two new arms, one on each side of the curve.

A typical start position has n vertices and no edges.

CHERRIES

Position: A row of red, blue, or green cherries.

Moves: Left removes a blue or green cherry from either end of the row; Right removes a red or green cherry similarly.

Variants: RED-BLUE CHERRIES has no green-colored cherries.

CLEAR THE POND

Position: A finite strip of squares, each square is empty or occupied by a blue or red piece.

Moves: Left moves a blue piece to the right to the next empty space or off the board if there is no empty space. Right moves leftward.

Variants: In the ALL-SMALL variant, the game component is over when only pieces of one color remain. See also CLOBBER THE POND.

CLOBBER

Position: A rectangular board, each square is empty or occupied by a blue or red stone. Play often begins with a board filled with alternating blue and red stones in a checkerboard pattern.

Moves: Left moves a blue stone onto an orthogonally adjacent red stone and removes the red stone. Right moves a red stone onto an orthogonally adjacent blue stone and removes the blue stone.

CLOBBER THE POND

Position: A finite strip of squares, each square is empty or occupied by a blue or red piece.

Moves: Left moves a blue piece rightward to the next empty space or off the board if there is no empty space. All pieces (of either color) that this piece jumps over are removed. Right moves leftward.

Variants: CLEAR THE POND

COL

See SNORT.

CRAM

See DOMINEERING.

CRICKET PITCH

Position: A sequence of bumps (positive integers) and one roller.

Moves: Left moves the roller to the right any number of bumps. Each rolled bump is reduced by 1. A value of 0 is no longer a bump, and can no longer be rolled over.

Note that once the roller reduces a bump to 0, parts of the lawn are no longer accessible to the roller and so can be removed as in Right's first (and worst) move below:

$$3, \odot, 6, 5, 1, 2, 4 \xrightarrow{L} \odot, 2, 6, 5, 1, 2, 4 \xrightarrow{R} 1, \odot, 4 \xrightarrow{L} \odot$$

Variants: In ROLL THE LAWN, the roller is allowed to go over non-bumps (zeros), so long as at least one bump is rolled.

CUTTHROAT

Position: A graph in which each vertex is colored red or blue.

Moves: Left removes a blue vertex. When a vertex is removed, all the incident edges and all monochromatic connected components are also removed.

Let $\bullet\!\!<^m_n$ ($\circ\!\!<^m_n$), $m+n \geq 1$, be a star graph with a blue (red) center, m blue leaves, and n red leaves.

$$\bullet\!\!<^2_2 \xrightarrow{L} \bullet\!\!<^2_1 \xrightarrow{R} \bullet\!\!<^1_1 \xrightarrow{L} \bullet\!\!<^1_0 = \bullet\!\!<^1$$

Variants: CUTTHROAT STARS is CUTTHROAT restricted to a collection of star graphs.

DOMINEERING

Position: A subset of the squares of a grid. Play often begins with a complete $n \times n$ square.

Moves: Left places a domino to cover/remove two adjacent vertical squares. Right places horizontally.

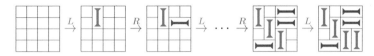

Variants: CRAM is impartial; either player can play a domino vertically or horizontally.

DOTS & BOXES

Position: A grid of dots, with lines drawn connecting some pairs of orthogonally adjacent dots. Any 1×1 completed square box of lines contains a player's initial. Play often begins with an empty rectangular grid of dots.

Moves: A move consists of connecting two adjacent dots by a line. If that move completes one or two boxes, the player initials the box(es) and must move again unless all boxes have been claimed.

Winner: The game ends when all boxes are claimed, and the player who has claimed the most boxes wins.

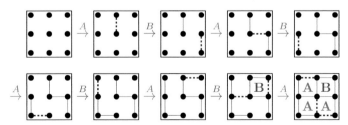

DRAG OVER

Position: A finite strip of squares, each square is empty or occupied by a blue or red piece.

Moves: A piece of the current player's color is moved one square to the left replacing any piece that may be on that square. All pieces to the right of the piece moved are also moved one square to the left. Pieces can be moved off the end of the strip.

Variants: In ALL-SMALL DRAG OVER the game is over when only pieces of one color remain on the board. See also RUN OVER, SHOVE, and PUSH.

ELEPHANTS & RHINOS

Position: A finite strip of squares, each square is empty or occupied by a blue (elephant) or red (rhino) piece.

Moves: Left moves an elephant one square to the right onto an empty square. No beast can budge or land on another. Right moves left-ward.

ENDNIM

Position: A row of stacks of boxes.

Moves: A player chooses one end of the row and removes any number of boxes from the stack at that end. Once a stack is empty, the next stack is available.

$$5, 3, 4 \xrightarrow{L} 1, 3, 4 \xrightarrow{R} 1, 3, 1 \xrightarrow{L} 1, 3 \xrightarrow{R} 1, 1 \xrightarrow{L} 1 \xrightarrow{R} 0$$

Variants: In PARTIZAN ENDNIM, Left can only remove boxes from the left end and Right from the right end.

EROSION

Position: A heap that contains l left counters and r right counters.

Moves: If $l > r$, Left may legally move by removing exactly r left counters. Similarly, if $r > l$, Right may remove l right counters.

The heap is denoted (l, r).

$$(14, 9) \xrightarrow{L} (5, 9) \xrightarrow{R} (5, 4) \xrightarrow{L} (1, 4) \xrightarrow{R} (1, 3)$$

FORKLIFT

Position: A row of stacks of boxes.

Moves: A forklift can pick up any number of boxes from an end stack. Left takes from the left end and Right from the right end; each puts them on the next stack in the row provided that what remains of the first stack is lower than the height of the second stack (so that the forklift can put the boxes down). The game is finished when there is only one stack.

$$2,1,3,2 \overset{L}{\to} 3,3,2 \overset{R}{\to} 3,4,1 \overset{L}{\to} 1,6,1 \overset{R}{\to} 1,7 \overset{L}{\to} 8$$

Variants: In IMPARTIAL FORKLIFT players can take from either end.

GEOGRAPHY

Position: A directed graph G with a designated vertex v. Play often begins with G being a graph whose vertices are represented by place names, and where there is an edge from p to q if the final letter of p is the initial letter of q.

Moves: An out-neighbor of v is designated, and v is deleted from the graph. This can be thought of as constructing a directed walk in the original graph that is not allowed to revisit any vertex.

$$\text{England} \overset{A}{\to} \text{Denmark} \overset{B}{\to} \text{Kyrgyzstan} \overset{A}{\to} \cdots$$

HACKENBUSH

Position: A graph with edges colored blue, red, and green. There is one special vertex called the *ground,* which is shown by a long horizontal line.

Moves: Left cuts a blue or green edge and removes any portion of the graph no longer connected to the ground.

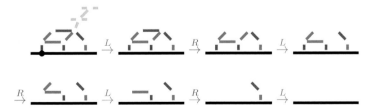

Variants: LR-HACKENBUSH positions have no green edges. GREEN (or IMPARTIAL) HACKENBUSH positions have only green edges.

HEX

Position: An $n \times n$ hexagonal tiling possibly containing some blue and red stones. Play usually begins with an empty board (but see the note in the variants).

Moves: Blue and Red alternate placing stones on empty hexagons.

Winner: Blue hopes to connect the upper-left side to the lower right with a path of her color, while Right tries to connect the lower left to the upper right.

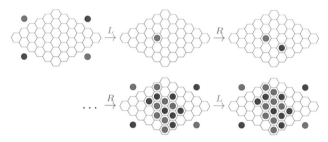

Variants: In practice, particularly on a smaller board, this is a game best played with the *swap rule* to negate the first-player advantage. Alice places a stone (say a blue stone), and then Bob decides whether to continue with red or whether he wants to swap colors and play blue. If he swaps, Alice plays again, this time placing a red stone. The game SQUEX is like HEX but is played on a square board and, unlike HEX, can end in a draw.

KAYLES

Position: A row of pins, possibly containing some gaps.

Moves: Throw a ball that removes either one or two adjacent pins.

Variants: GENERALIZED KAYLES or OCTAL games where a heap can be reduced in size and then split into two heaps.

KONANE

Position: A rectangular board, each square is empty or occupied by a blue or red stone. Play often begins with a square or rectangular board filled with alternating blue and red stones in a checkerboard pattern with two adjacent stones removed from central squares.

Moves: Left jumps a blue stone over an orthogonally adjacent red stone onto an empty square, removing the red stone that was jumped over. In a single move, jumps of this type *may* be chained together, but only in a single direction; that is, the jumping stone may not make a 90° turn.

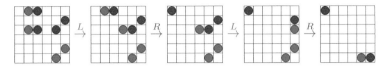

MAZE

Position: A rectangular grid oriented so that the edges are at 45 degree angles to the horizontal, with some highlighted edges and one piece (or multiple pieces) on some square(s).

Moves: Left moves the piece any number of squares in a southwesterly direction; Right moves the piece similarly in a southeasterly direction. The piece may not cross a highlighted edge. If multiple pieces are on the board, they do not interfere with one another, making the position a disjunctive sum.

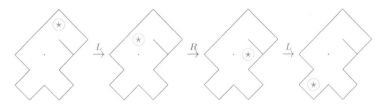

Variants: In MAIZE each player can only move the piece one square. This piece is denoted ⊙.

NIM

Position: Several heaps of counters.

Moves: A player chooses a heap and removes some counters.

$$9, 4, 2 \xrightarrow{A} 3, 4, 2 \xrightarrow{B} 3, 4, 1 \xrightarrow{A} 3, 2, 1 \xrightarrow{B} 0, 2, 1 \xrightarrow{A} 0, 0, 1 \xrightarrow{B} 0, 0, 0$$

Variants: In SUBTRACTION games and ALLBUT subtraction games, the number of counters that a player can remove is restricted in some way; in GREEDY NIM a player must take from the largest heap. In POKER NIM, a player can add to a heap using counters taken early in the game.

PUSH

> Position: A finite strip of squares, each square is empty or occupied by a blue or red piece. (The pieces for PUSH look like those for SHOVE, but with fewer lines behind the hand.)

> Moves: A piece of the current player's color is pushed one square to the left. No square can be occupied by more than one piece, so any pieces immediately to the left of this piece are also pushed one square. Pieces may be pushed off the end of the strip.

> Variants: In ALL-SMALL PUSH the game is over when only pieces of one color remain on the board. See also SHOVE.

ROLL THE LAWN

> Position: A sequence of bumps (positive integers) and non-bumps (zeros) and one roller.

> Moves: Left moves the roller to the right any number of bumps. Each rolled bump is reduced by 1. A value of 0 is no longer a bump and has no further effect on the game.

$$3, \circ, 1, 1, 2 \overset{L}{\to} \circ, 2, 1, 1, 2 \overset{R}{\to} 1, 0, \circ, 1, 2$$
$$\overset{L}{\to} \circ, 0, 0, 1, 2 \overset{R}{\to} 0, 0, 0, \circ, 2$$

> Variants: In CRICKET PITCH, the roller is prohibited from rolling over non-bumps (zeros).

RUN OVER

> Position: A finite strip of squares, each square is empty or occupied by a blue or red piece.

> Moves: A piece of the current player's color is moved one square to the left replacing any piece that may be on that square. Pieces may be moved off the end of the strip.

> Variants: In ALL-SMALL RUN OVER the game is over when only pieces of one color remain on the board. See also DRAG OVER, SHOVE, and PUSH.

SHE LOVES ME SHE LOVES ME NOT

Position: A trivial children's game played on a daisy with n petals.

Moves: Players alternate removing a petal until none are remaining. Naturally, only the parity of the number of petals matters.

SHOVE

Position: A finite strip of squares, each square is empty or occupied by a blue or red piece. (The pieces for SHOVE look like those for PUSH, but with more lines behind the hand.)

Moves: A piece of the current player's color and all pieces to its left are pushed one square to the left. Pieces may be shoved off the end of the strip.

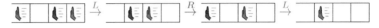

Variants: In ALL-SMALL SHOVE the game is over when only pieces of one color remain on the board. See also PUSH.

SNORT

Position: A graph with each vertex either uncolored or colored red or blue.

Moves: Left colors an uncolored vertex blue, subject to the proviso that the vertex being colored is not adjacent to a red vertex.

Variants: In COL, Left is not allowed to color a vertex adjacent to a blue vertex.

SPLITTLES$(L \mid R)$

Position: A heap of n counters.

Moves: Left subtracts some element of set L from n and then (optionally) splits the remaining heap into two heaps.

If $L = \{2, 3, 7\}$, $R = \{2, 4\}$, and $n = 20$,

$$20 \xrightarrow{L} \{9, 8\} \xrightarrow{R} \{3, 4, 8\} \xrightarrow{L} \{3, 4, 6\} \xrightarrow{R} \{3, 6\} \xrightarrow{L} \{3, 2\} \xrightarrow{R} \{1, 2\} \xrightarrow{L} \{1\}$$

Variants: Denote the impartial game SPLITTLES($S \mid S$) by SPLITTLES(S). See also SUBTRACTION($L \mid R$).

SPROUTS

Position: A collection of vertices in the plane connected by non-crossing edges. The maximum degree of a vertex is always 3. Play often begins with n vertices and no edges.

Moves: Draw a curve joining one vertex to another (or itself) that does not cross any other curve or pass through any other vertex. On this curve place another vertex. (Since the maximum-allowed degree is 3, the curve must join two vertices of degree at most 2, or a vertex of degree 0 or 1 to itself.)

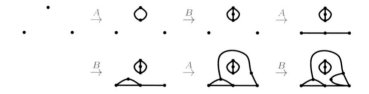

Variants: In SPROUTLETTES, the maximum degree of a vertex is 2 rather than 3. See also BRUSSELS SPROUTS.

STIRLING-SHAVE

Position: A sequence of positive-sized piles.

Moves: A player is allowed to choose any pile that is smaller than all piles to its right. The player then removes the chosen pile as well as all piles to its right.

$$[3, 4, 1, 5, 6] \xrightarrow{A} [3, 4, 1] \xrightarrow{B} [3, 4] \xrightarrow{A} \emptyset.$$

STRINGS & COINS

Position: A grid of coins, and a set of strings with one or both ends connected to a coin.

Moves: Cut (i.e., entirely remove) a string. If, after the move, one or two coin(s) have no strings, the player pockets the coin(s) and, if moves remain, must move again.

Winner: The player who pocketed the most coins.

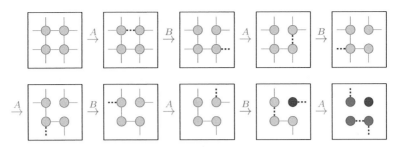

SUBTRACTION$(L \mid R)$

Position: A non-negative integer n.

Moves: Left subtracts some element of set L from n. Right subtracts an element of R from n. The result of the subtraction must be non-negative.

If $L = \{2, 3\}$, $R = \{1, 4, 5\}$, and $n = 17$,

$$17 \xrightarrow{L} 14 \xrightarrow{R} 13 \xrightarrow{L} 10 \xrightarrow{R} 6 \xrightarrow{L} 3 \xrightarrow{R} 2 \xrightarrow{L} 0$$

Variants: With one subtraction set, SUBTRACTION(S) is the impartial game SUBTRACTION$(S \mid S)$. In the literature, SUBTRACTION(S) is the standard *subtraction game,* and SUBTRACTION$(L \mid R)$ is termed a *partizan subtraction game.* The game ALLBUT(S) is the game SUBTRACTION$(\{1, 2, \ldots\} \setminus S)$; S should be a finite set.

TIMBER

Position: A row of dominoes, each askew toward the left or right.

Moves: Either player can tip a domino in the direction it's already tilted, toppling over all dominoes in its path.

$$\text{▮▮▮} \xrightarrow{A} \text{▮▮} \xrightarrow{B} \underline{}$$

Variants: TOPPLING TOWERS is the two-dimensional version played on a checkerboard. If a tower topples into an empty space, then the "chain reaction" stops. The game can be extended into any number of dimensions.

TOPPLING DOMINOES

Position: A row of blue, red, or green dominoes.

Moves: Left chooses a blue or green domino and topples it either left or right; every domino in that direction also topples and is removed from the game. Right topples a red or green domino.

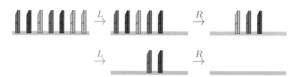

Variants: TOPPLING TOWERS is the two-dimensional version played on a checkerboard. If a tower topples into an empty space, then the "chain reaction" stops. The game can be extended into any number of dimensions.

WOODPUSH

Position: A finite strip of squares, each square is empty or occupied by a blue or red piece.

Moves: A player may either *push* or *retreat*. (Left pushes rightward, Right pushes leftward, and retreats are in the opposite direction.) A retreat involves moving one square backward (or off the end) and is legal only if that square is empty. A push moves a continuous row of pieces ahead (as in PUSH), but the move *must* include a piece of the opposing color. Repetition of global board positions is not allowed; (i.e., a *ko* threat must be played first).

Bibliography

Page numbers in *italics* after a bibliography entry refer to the locations of citations within *this* text.

[AGNW05]   Michael H. Albert, J. P. Grossman, Richard J. Nowakowski, and David Wolfe. An introduction to clobber. *Integers: Electronic Journal of Combinatorial Number Theory*, 5(2):A1, 2005. *252, 275.*

[AN01]   Michael Albert and Richard Nowakowski. The game of End-Nim. *The Electronic Journal of Combinatorics*, 8(2):R1, 2001. *51.*

[AN07]   Michael Albert and Richard Nowakowski, editors. *More Games of No Chance 3.* Number 56 in Mathematical Sciences Research Institute Publications. Cambridge University Press, Cambridge, UK, 2007. *xii, 318, 319, 320.*

[Aus76]   R. B. Austin. Impartial and partizan games. Master's thesis, University of Calgary, 1976. *175.*

[BCG01]   Elwyn R. Berlekamp, John H. Conway, and Richard K. Guy. *Winning Ways for Your Mathematical Plays.* A K Peters, Ltd., Natick, MA, second edition, 2001. First edition published in 1982 by Academic Press. *4, 6, 20, 74, 77, 101, 104, 154, 163, 175, 180, 196, 197, 206, 210, 211, 221, 222, 241, 251, 255.*

[Bec06]   József Beck. *Combinatorial Games: Tic-Tac-Toe Theory.* Number 114 in Encyclopedia of Mathematics and Its Applications. Cambridge University Press, Cambridge, UK, 2006. *61.*

[Ber88]   Elwyn R. Berlekamp. Blockbusting and domineering. *Journal of Combinatorial Theory*, 49(1):67–116, September 1988. *68.*

[Ber96]   Elwyn Berlekamp. The economist's view of combinatorial games. In Nowakowski [Now96], pages 365–404. *196, 255.*

[Ber00]   Elwyn Berlekamp. *The Dots and Boxes Game: Sophisticated Child's Play.* A K Peters, Ltd., Natick, MA, 2000. *26.*

[Ber02]   Elwyn Berlekamp. Idempotents among partisan games. In Nowakowski [Now02], pages 3–23. *255.*

[Bou02]   C. L. Bouton. Nim, a game with a complete mathematical theory. *Annals of Mathematics*, 3(2):35–39, 1902. *42, 59, 156.*

[BW94]   Elwyn Berlekamp and David Wolfe. *Mathematical Go: Chilling Gets the Last Point.* A K Peters, Ltd., Wellesley, MA, 1994. *198.*

317

[CD73]   Peter Crawley and Robert P. Dilworth. *Algebraic Theory of Lattices.*
           Prentice-Hall, Inc., Englewood Cliffs, NJ, 1973. *150.*

[Con01]   John H. Conway. *On Numbers and Games.* A K Peters, Ltd., Natick,
           MA, second edition, 2001. First edition published in 1976 by Academic
           Press. *6, 77, 101, 108, 137, 161, 180, 188, 211, 239, 241, 247, 252.*

[Con02]   John H. Conway. More infinite games. In Nowakowski [Now02], pages
           31–36. *252.*

[DDE02]   Eri D. Demaine, Martin L. Demaine, and David Eppstein. Phutball
           endgames are hard. In Nowakowski [Now02], pages 351–360. *252.*

[DKW03]   Adam Duffy, Garrett Kolpin, and David Wolfe. Ordinal partizan end
           nim. In Albert and Nowakowski [AN07], pages 419–425. *51.*

[DRSS15]   Paul Dorbec, Gabriel Renault, Aaron N. Siegel, and Éric Sopena. Di-
           cots, and a taxonomic ranking for misère games. *Journal of Combina-
           torial Theory*, 130:42–63, February 2015. *256.*

[Elk96]   Noam D. Elkies. On numbers and endgames: Combinatorial game the-
           ory in chess endgames. In Nowakowski [Now96], pages 135–150. *104.*

[Ett96]   J. Mark Ettinger. *Topics in Combinatorial Games.* PhD thesis, Univer-
           sity of Wisconsin, Madison, 1996. *257.*

[Ett00]   J. Mark Ettinger. A metric for positional games. *Theoretical Computer
           Science*, 230:207–219, 2000. *257.*

[FK87]   A. S. Fraenkel and A. Kotzig. Partizan octal games: Partizan subtrac-
           tion games. *International Journal of Game Theory*, 16:145–154, 1987.
           34.

[FN04]   R. Fleischer and R. J. Nowakowski, editors. *Algorithmic Combinatorial
           Game Theory, Dagstuhl, Germany, February 2002*, volume 313(3):313–
           546 of *Theoretical Computer Science*, February 2004. *xii.*

[FNSW15]   Alex Fink, Richard J. Nowakowski, Aaron N. Siegel, and David Wolfe.
           Toppling conjectures. In Nowakowski [Now15], pages 65–76. *246, 247,
           248.*

[Fra00]   Aviezri S. Fraenkel. Recent results and questions in combinatorial game
           complexities. *Theoretical Computer Science*, 249:265–288, 2000. Con-
           ference version in C. S. Iliopoulos, editor, *Proceedigs of AWOCA98 —
           Ninth Australasian Workshop on Combinatorial Algorithms*, pages 124–
           146, Perth, Western Australia, Curtin University, 1998. *253.*

[Fra04]   Aviezri Fraenkel. Complexity, appeal and challenges of combinatorial
           games. *Theoretical Computer Science*, 313(3):393–415, February 2004.
           253.

[Grä71]   George Grätzer. *Lattice Theory: First Concept and Distributive Lat-
           tices.* W. H. Freeman and Company, San Francisco, 1971. *150.*

[Gru39]   P. M. Grundy. Mathematics and games. *Eureka*, 2:6–8, 1939. Reprinted
           in *Eureka* 27:9–11, 1964. *59, 156.*

[GS56]   Richard K. Guy and Cedric A. B. Smith. The *G*-values of various games.
           Proceedings of the Cambridge Philosophical Society, 52:514–526, 1956.
           156.

[GS07]   J. P. Grossman and Aaron N. Siegel. Reductions of partizan games. In
            Albert and Nowakowski [AN07], pages 427–445. *234, 237*.

[Guy89]   Richard K. Guy. *Fair Game: How to Play Impartial Combinatorial
            Games*. COMAP Mathethematical Exploration Series. COMAP, Inc.,
            Arlington, MA, 1989. *156*.

[Guy91]   Richard K. Guy, editor. *Combinatorial Games*, volume 43 of *Proceedings
            of Symposia in Applied Mathematics*. American Mathematical Society,
            Providence, RI, 1991. Lecture notes prepared for the American Math-
            ematical Society Short Course held in Columbus, Ohio, August 6–7,
            1990. *xii*.

[HN03]   D. G. Horrocks and R. J. Nowakowski. Regularity in the $\mathcal{G}$-sequences of
            octal games with a pass. *Integers: Electronic Journal of Combinatorial
            Number Theory*, 3:G1, 2003. *164*.

[HN04]   S. Howse and R. J. Nowakowski. Periodicity and arithmetic-periodicity
            in hexadecimal games. *Theoretical Computer Science*, 313(3):463–472,
            2004. *164*.

[Knu74]   Donald E. Knuth. *Surreal Numbers*. Addison-Wesley, Reading, MA,
            1974. *77, 108, 252*.

[LaPNNS16] Urban Larsson, Jo ao P. Neto, Richard J. Nowakowski, and Carlos P.
            Santos. Guaranteed scoring games. *The Electronic Journal of Combi-
            natorics*, 23(3):P3.27, 2016. *257*.

[Lev06]   Lionel Levine. Fractal sequences and restricted nim. *Ars Combinatoria*,
            80:113–127, 2006. *178*.

[LMR96]   M. Lachmann, C. Moore, and I. Rapaport. Who wins domineering on
            rectangular boards? In Nowakowski [Now96], pages 307–315. *68*.

[McK16]   Neil A. McKay. *Forms and Values of Number-Like and Nimber-Like
            Games*. PhD thesis, Dalhousie University, 2016. *239*.

[ML06]   M. Müller and Z. Li. Locally informed global search for sums of combina-
            torial games. In *Computers and Games: 4th International Conference,
            CG 2004, Ramat-Gan, Israel, July 5–7, 2004, Revised Papers*, number
            3846 in Lecture Notes in Computer Science, pages 273–24. Springer-
            Verlag, New York, 2006. *196*.

[Moe91]   David Moews. Sums of games born on days 2 and 3. *Theoretical Com-
            puter Science*, 91(1):119–128, 1991. *151*.

[MR12]   Rebecca Milley and Gabriel Renault.   Dead ends in misère play:
            The misère monoid of canonical numbers.   *Discrete Mathematics*,
            313(20):2223–2231, December 2012. *256*.

[NO11]   Richard J. Nowakowski and Paul Ottaway. Option-closed games. *Con-
            tributions to Discrete Mathematics*, 6(1):142–153, 2011. *96, 235*.

[Now96]   Richard Nowakowski, editor.   *Games of No Chance: Combinatorial
            Games at MSRI, 1994*. Number 29 in Mathematical Sciences Research
            Institute Publications. Cambridge University Press, 1996. *xii, 317, 318,
            319*.

[Now02]   Richard Nowakowski, editor. *More Games of No Chance*. Number 42 in Mathematical Sciences Research Institute Publications. Cambridge University Press, Cambridge, UK, 2002. *xii, 317, 318, 320*.

[Now15]   Richard Nowakowski, editor. *More Games of No Chance 4*. Number 63 in Mathematical Sciences Research Institute Publications. Cambridge University Press, Cambridge, UK, 2015. *xii, 318, 320*.

[SA03]   Svetoslav Savchev and Titu Andreescu. *Mathematical Miniatures*. Number 43 in New Mathematical Library. MAA Press, Washington, DC, 2003. *18*.

[Sie03]   Aaron N. Siegel. CGSuite or combinatorial game suite. Software toolkit for analyzing combinatorial games. http://www.cgsuite.org, 2003. *271*.

[Sie05]   Aaron N. Siegel. *Loopy Games and Computation*. PhD thesis, University of California, Berkeley, Spring 2005. *77, 254*.

[Sie06]   Angela Siegel. Finite excluded subtraction sets and infinite modular nim. Master's thesis, Dalhousie University, 2006. *173*.

[Sie07]   Aaron N. Siegel. New results in loopy games. In Albert and Nowakowski [AN07], pages 215–232. *254*.

[Sie11]   Angela Siegel. *On the Structure of Games and Their Posets*. PhD thesis, Dalhousie University, 2011. *96*.

[Sie13]   Aaron N. Siegel. *Combinatorial Game Theory*. American Mathematical Society, Washington, DC, 2013. *234, 237, 240, 251, 252, 254, 255, 256*.

[Sie15]   Aaron N. Siegel. Misère canonical forms of partizan games. In Nowakowski [Now15], pages 225–240. *257*.

[Slo]   N. J. A. Sloane. The on-line encyclopedia of integer sequences (OEIS). https://oeis.org. *49*.

[SP08]   Aaron N. Siegel and Thane E. Plambeck. Misère quotients for impartial games. *Journal of Combinatorial Number Theory, Series A*, 115(4):593–622, May 2008. Supplement at https://arxiv.org/abs/0705.2404. *256*.

[Spi01]   William L. Spight. Extended thermography for multiple kos in Go. *Theoretical Computer Science*, 252(1-2):23–43, 2001. *254*.

[Spi02]   William L. Spight. Go thermography: the 4/21/98 Jiang-Rui endgame. In Nowakowski [Now02], pages 89–105. *255*.

[Spr35]   R. P. Sprague. Üeber mathematische Kampfspiele. *Tôhoku Mathematics Journal*, 13(290):438–444, 1935. *59, 156*.

[Wes01]   Douglas B. West. *Introduction to Graph Theory*. Prentice Hall, Upper Saddle River, NJ, second edition, 2001. *20*.

[WF04]   David Wolfe and William Fraser. Counting the number of games. *Theoretical Computer Science*, 313(3):527–532, February 2004. *141*.

[Wol93]   David Wolfe. Snakes in domineering games. *Theoretical Computer Science*, 119(2):323–329, October 1993. *200*.

[Wol02]   David Wolfe. Go endgames are PSPACE-hard. In Nowakowski [Now02], pages 125–136. *125*.

[YB82]   V. K. Yu and R. B. Banerji. Periodicity of sprague-grundy function in graphs with decomposable nodes. *Cybernetics and Systems: An International Journal*, 13(4):299–310, 1982. *176*.

Index

In the index, page numbers in *italics* refer to a definition, ruleset, or theorem.